LONDON MATHEMATICAL SOCIETY LECTURE NOTE SERIES

Managing Editor: Professor I.M.James,
Mathematical Institute, 24-29 St Giles, Oxford

T0269354

1. General cohomology theory and K-theory, P.HILTON
4. Algebraic topology: a student's guide, J.F.ADAMS
5. Commutative algebra, J.T.KNIGHT
8. Integration and harmonic analysis on compact groups, R.E.EDWARDS
9. Elliptic functions and elliptic curves, P.DU VAL
10. Numerical ranges II, F.F.BONSALL & J.DUNCAN
11. New developments in topology, G.SEGAL (ed.)
12. Symposium on complex analysis, Canterbury, 1973, J.CLUNIE & W.K.HAYMAN (eds.)
13. Combinatorics: Proceedings of the British combinatorial conference 1973, T.P.McDONOUGH & V.C.MAVRON (eds.)
14. Analytic theory of abelian varieties, H.P.F.SWINNERTON-DYER
15. An introduction to topological groups, P.J.HIGGINS
16. Topics in finite groups, T.M.GAGEN
17. Differentiable germs and catastrophes, Th.BROCKER & L.LANDER
18. A geometric approach to homology theory, S.BUONCRISTIANO, C.P.ROURKE & B.J.SANDERSON
20. Sheaf theory, B.R.TENNISON
21. Automatic continuity of linear operators, A.M.SINCLAIR
23. Parallelisms of complete designs, P.J.CAMERON
24. The topology of Stiefel manifolds, I.M.JAMES
25. Lie groups and compact groups, J.F.PRICE
26. Transformation groups: Proceedings of the conference in the University of Newcastle upon Tyne, August 1976, C.KOSNIOWSKI
27. Skew field constructions, P.M.COHN
28. Brownian motion, Hardy spaces and bounded mean oscillation, K.E.PETERSEN
29. Pontryagin duality and the structure of locally compact abelian groups, S.A.MORRIS
30. Interaction models, N.L.BIGGS
31. Continuous crossed products and type III von Neumann algebras, A.VAN DAELE
32. Uniform algebras and Jensen measures, T.W.GAMELIN
33. Permutation groups and combinatorial structures, N.L.BIGGS & A.T.WHITE
34. Representation theory of Lie groups, M.F.ATIYAH et al.
35. Trace ideals and their applications, B.SIMON
36. Homological group theory, C.T.C.WALL (ed.)
37. Partially ordered rings and semi-algebraic geometry, G.W.BRUMFIEL
38. Surveys in combinatorics, B.BOLLOBAS (ed.)
39. Affine sets and affine groups, D.G.NORTHCOTT
40. Introduction to H_p spaces, P.J.KOOSIS
41. Theory and applications of Hopf bifurcation, B.D.HASSARD, N.D.KAZARINOFF & Y-H.WAN
42. Topics in the theory of group presentations, D.L.JOHNSON
43. Graphs, codes and designs, P.J.CAMERON & J.H.VAN LINT
44. Z/2-homotopy theory, M.C.CRABB
45. Recursion theory: its generalisations and applications, F.R.DRAKE & S.S.WAINER (eds.)
46. p-adic analysis: a short course on recent work, N.KOBLITZ
47. Coding the Universe, A. BELLER, R. JENSEN & P. WELCH
48. Low-dimensional topology, R. BROWN & T.L. THICKSTUN (eds.)
49. Finite geometries and designs, P. CAMERON, J.W.P. HIRSCHFELD & D.R. HUGHES (eds.)
50. Commutator Calculus and groups of homotopy classes, H.J. BAUES
51. Synthetic differential geometry, A. KOCK
52. Combinatorics, H.N.V. TEMPERLEY (ed.)
53. Singularity theory, V.I. ARNOLD
54. Markov processes and related problems of analysis, E.B. DYNKIN
55. Ordered permutation groups, A.M.W. GLASS
56. Journees arithmetiques 1980, J.V. ARMITAGE (ed.)
57. Techniques of geometric topology, R.A. FENN
58. Singularities of differentiable functions, J. MARTINET
59. Applicable differential geometry, F.A.E. PIRANI and M. CRAMPIN
60. Integrable systems, S.P. NOVIKOV et al.

London Mathematical Society Lecture Note Series. 50

Commutator Calculus and Groups of Homotopy Classes

Hans Joachim Baues

Math. Institut der Universität Bonn
Sonderforschungsbereich 40 'Theoretische Mathematik'
Wegelerstr. 10
53 Bonn

CAMBRIDGE UNIVERSITY PRESS
CAMBRIDGE
LONDON NEW YORK NEW ROCHELLE
MELBOURNE SYDNEY

CAMBRIDGE UNIVERSITY PRESS
Cambridge, New York, Melbourne, Madrid, Cape Town, Singapore, São Paulo

Cambridge University Press
The Edinburgh Building, Cambridge CB2 8RU, UK

Published in the United States of America by Cambridge University Press, New York

www.cambridge.org
Information on this title: www.cambridge.org/9780521284240

First published 1981
Re-issued in this digitally printed version 2008

A catalogue record for this publication is available from the British Library

Library of Congress Catalogue Card Number: 81–10142

ISBN 978-0-521-28424-0 paperback

CONTENTS

page

Introduction to Part B 1

Introduction to Part A 9

Part A: Homotopy operations, nilpotent group theory and nilpotent Lie algebra theory

I. Commutator calculus 14

I § 1 The exponential function and the Zassenhaus formula 14

I § 2 The exponential commutator 22

I § 3 A presentation for the exponential group 26

I § 4 The general type of Zassenhaus terms and its characterization modulo a prime 29

II. Distributivity laws in homotopy theory 35

II § 1 Whitehead products and cup products 36

II § 2 Hopf invariants 42

II § 3 The Whitehead product of composition elements 49

II § 4 Proof of I (1.13) and I (2.6) 57

II § 5 Decomposition of suspensions and groups of homotopy classes 61

III. Homotopy operations on spheres 68

III § 1 Spherical Whitehead products and commutators 68

III § 2 Spherical Hopf invariants 71

III § 3 Deviation from commutativity of spherical cup products 75

III § 4 Cup products of spherical Hopf invariants 78

III § 5 Hopf invariants of a Hopf invariant, of a sum and of a cup product 83

III § 6 Hopf invariant of a composition element 85

IV. Higher order Hopf invariants on spheres 91

IV § 1 Fxamples of higher order Hopf invariants on spheres 91

IV § 2 Proof of theorem (1. 3) 94

IV § 3 Zassenhaus terms for an odd prime 98

Part B: Homotopy theory over a subring R of the rationals $\mathbb{Q}$ with 1/2, 1/3 $\in$ R 101

V. The homotopy Lie algebra and the spherical cohomotopy algebra 101

V § 0 Notation 101

V § 1 The homotopy Lie algebra and the spherical cohomotopy algebra 103

V § 2 Homotopy groups of spheres and homotopy coefficients 108

V § 3 The Hurewicz and the degree map 113

VI. Groups of homotopy classes 122

VI § 1 Nilpotent rational groups of homotopy classes 122

VI § 2 The exponential group 124

VI § 3 Groups of homotopy classes 126

VI § 4 H-maps and Co-H-maps 131

VII. The Hilton-Milnor theorem and its dual 135

VII § 1 The category of coefficients 135

VII § 2 Extension of algebras by homotopy coefficients 139

VII § 3 The extension of Lie algebras by homotopy coefficients 144

VII § 4 The Hilton-Milnor theorem and its dual 150

Literature 156

Index 159

INTRODUCTION TO PART B

An inexhaustible source of algebraic topology is the homotopy classification problem. If we have a space we would like to know a list of algebraic invariants which determine the homotopy type of the space. If we have a map we would like to characterize the map up to homotopy by algebraic invariants. Moreover, if the set of homotopy classes [X, G] is a group, for example if G is a topological group, we would like to determine the group structure of the set as well. Helpful tools for these problems are functors from the topological category to an algebraic category like homology, cohomology, homotopy groups etc. However, the known functors give us only rather crude algebraic pictures and almost nothing is known about the image categories of these functors.

There are two opposite directions in which the problem can be pursued, namely rational homotopy theory and stable homotopy theory. Both are studied with great energy. Indeed, research camps seem to have formed - on one side of the front are those mathematicians who think a rational space is the most natural object, on the other side those for whom a spectrum is the most natural object to start with. At the time of J. H. C. Whitehead people were interested in finite polyhedra. Soon they realized that the calculation of homotopy groups of spheres is a deep and fundamental obstacle to solving the classification problems. Rationally these groups were computed by Serre. Modulo a prime there are partial results, but the nature of the groups is essentially still unknown. Moreover, we have a complete rational solution of the homotopy classification problem in the results of Quillen and Sullivan. They have shown that the homotopy categories of differential graded Lie algebras over $\mathbb{Q}$ and of differential graded commutative algebras over $\mathbb{Q}$ are equivalent to the homotopy category of 1-connected rational spaces. In view of this we may now well doubt whether stable homotopy theory is a 'first approximation to homotopy theory' as it was conceived to be by J. H. C. Whitehead and Spanier (in Proc. Nat. Acad. Sci. 39 (1953), 655-60). Instead, we might try to understand the homotopy classification problem by extending the rational solution of the problem to a solution over a subring R of the rationals. It is one purpose of this book to initiate such an investigation.

The only result in the literature which illustrates our approach is the Hilton-Milnor theorem on the graded homotopy group of a one-point union of spheres. Rationally this group is just a free graded Lie algebra. Hilton showed that over the integers we still have a direct sum decomposition of these groups in terms of a basis in a free Lie algebra and in terms of homotopy groups of spheres. We believe in fact that over the integers this group should also be a free object in a suitable category. It developed that a description of even this category requires a formidable apparatus. However, for a subring R of the rationals $\mathbb{Q}$ with 1/2, $1/3 \in R$ a category as in the following problem has an elegant characterization. Let Top_ε be the homotopy category of ε-connected spaces.

(1) Problem. Construct a category Lie_M such that the functor of homotopy groups

$$L(\,.\,, R): Top_1 \to Lie_M \text{ with}$$

$$L(X, R) = \pi_*(\Omega X) \otimes R \quad \text{(endowed with suitable algebraic structure)}$$

maps a one-point union of spheres V to a free object L(V, R) in Lie_M.

Clearly Whitehead or Samelson products give the graded module $L(X, R) = \pi_*(\Omega X) \otimes R$ the structure of a graded Lie algebra over R. Only for $R = \mathbb{Q}$ do we know that L(V, Q) is also a free Lie algebra. For $R \neq \mathbb{Q}$ we have to introduce additional structure on L(X, R) to obtain the objects in the category Lie_M required by problem (1). At first sight this problem might appear to be a merely formal question. The solution nevertheless is essential to our notion of extending rational homotopy theory to a theory over a subring R of $\mathbb{Q}$. The category Lie_M should play a role in homotopy theory over R similar to that of the category $Lie_{\mathbb{Q}}$ of graded rational Lie algebras in rational homotopy theory. Clearly the non-rational theory is enormously more complicated than the rational one. Still, the study of the rational situation can serve as a guide. We will give various examples to illustrate this.

Rational homotopy theory can be developed in two ways dual to each other, namely via the cofibration or Lie algebra method of Quillen and by the fibration or commutative algebra method of Sullivan. It turned out that the cohomology functor $H^*(. , \mathbb{Q})$ has properties completely dual to those of the homotopy functor $L(. , \mathbb{Q})$. This leads to our next problem.

(2) Problem. Is there a category Div alg_M and a functor

$$M(. , R) : \text{Top}_o \to \text{Div alg}_M$$

with properties dual to those of the functor $L(. , R)$ so that this duality extends the known duality of the rational functors

$$M(. , \mathbb{Q}) = H^*(. , \mathbb{Q}) \text{ and}$$

$$L(. , \mathbb{Q}) = \pi_*(\Omega\, .) \otimes \mathbb{Q} \, ?$$

Clearly the formulation of this problem is not very precise. It expresses only our feeling of what it would be nice to have. We will show that there is such a functor $M(. , R)$, which we call the spherical cohomotopy functor. For a finite dimensional polyhedron X the graded R-module $M(X, R)$ is given by the set of homotopy classes

$$M^n(X, R) = [X, \Omega^n_R]$$

where for the R-local n-sphere S^n_R we set

$$\Omega^n_R = \begin{cases} S^n_R & n \text{ odd} \\ \Omega\Sigma S^n_R & n \text{ even} > 0 . \end{cases}$$

By a result of Serre we know that $\Omega^n_{\mathbb{Q}} = K(\mathbb{Q}, n)$ is an Eilenberg-MacLane space. It is well known that for a product $P_{\mathbb{Q}}$ of such Eilenberg-MacLane spaces the cohomology algebra $H^*(P_{\mathbb{Q}}, \mathbb{Q})$ is a free graded commutative algebra. More generally, we found that for a product P_R of spaces Ω^n_R the algebra $M(P_R, R)$ is also a free object in the category Div alg_M, which we construct. This category is appropriate for homotopy theory over R and generalizes the category of rational commutative algebras.

Although the construction of the categories Lie_M and $Div\ alg_M$ is quite intricate, it can be sketched as follows. The homotopy groups of spheres give us the double graded R-module $M = M_R$ with

$$M_R^{n,m} = M^n(S^m, R) = \begin{cases} \pi_m(S^n) \otimes R & n \text{ odd} \\ \pi_{m+1}(S^{n+1}) \otimes R & n \text{ even} . \end{cases}$$

This module with additional structure (namely smash product, higher-order Hopf invariants γ_p for each prime p and units $e^r \in M_R^{r,r} = R$) is an object in the category $Coef_R$ of coefficients. It turns out that there is an associative tensor product $\tilde{\otimes}$ in this category $Coef_R$ which we can use to define the notion of a monoid in $Coef_R$. We show that in fact the composition $\odot$ of maps gives M_R the monoid structure

$$\odot : M_R \tilde{\otimes} M_R \to M_R$$

in $Coef_R$. For the categories

$$\begin{cases} Lie_R & = \text{category of graded Lie algebras over } R \\ Div\ alg_R & = \text{category of graded commutative algebras over } R \\ & \quad \text{with divided powers} \end{cases}$$

we construct bifunctors

$$(3) \quad \begin{cases} Lie_R \times Coef_R & \overset{\tilde{\otimes}}{\to} & Lie_R \\ Coef_R \times Div\ alg_R & \overset{\tilde{\otimes}}{\to} & Div\ alg_R \end{cases}$$

which are associative with respect to the tensor product $\tilde{\otimes}$ in $Coef_R$. With these 'twisted' products $\tilde{\otimes}$ we define actions of the monoid M_R to be morphisms

$$(4) \quad \begin{cases} \odot : L \tilde{\otimes} M_R \to L & \text{in } Lie_R \\ \odot : M_R \tilde{\otimes} A \to A & \text{in } Div\ alg_R \end{cases}$$

which are associative with respect to the monoid structure $\odot$ on M_R.

The objects of Lie_M and $Divalg_M$ are now just the objects of Lie_R and $Divalg_R$ together with such an action. The morphisms are just the equivariant maps in Lie_R or $Divalg_R$ respectively.

In this way we replace the coefficient module $\mathbb{Q} = M_{\mathbb{Q}}$ of rational homotopy theory by the coefficient module $M = M_R$ needed in homotopy theory over R. Here we are not deterred by not knowing explicitly the homotopy groups of spheres $M_R^{n,m}$. We just clarify the 'primary' algebraic structure of these groups, namely their structure as a monoid in $Coef_R$. The construction of the categories Lie_M and $Divalg_M$ depends only on this primary structure. We will prove that there exist free objects in these categories. This now allows us to formulate the first basic classification result.

(5) Theorem. (A) The full subcategory of Top_1 of spaces homotopy equivalent to a one-point union of finitely many R-local spheres S_R^n is, via the functor $L(\,.\,, R)$, equivalent to the full subcategory of Lie_M of finitely generated free objects.

(B) The full subcategory of Top_0 of spaces homotopy equivalent to a product of finitely many spaces Ω_R^n is, via the functor $M(\,.\,, R)$, equivalent to the full subcategory of $Divalg_M$ of finitely generated free objects.

For $R = \mathbb{Q}$ this is a well known result of rational homotopy theory, in fact for $R = \mathbb{Q}$ it is the restriction of the equivalences of Sullivan and Quillen to the case of zero differentials.

We investigate the connection of the homotopy functors $L(\,.\,, R)$ and $M(\,.\,, R)$ with the corresponding homology and cohomology functors. That is, we consider the Hurewicz and degree maps, which are natural transformations

$$(6)\qquad \begin{cases} \Phi = \Phi_Y & : L(Y, R) \to PH_*(\Omega Y, R) \\ \deg = \deg_X & : M(X, R) \to H^*(X, R) \end{cases}$$

of Lie algebras and algebras respectively. Here we restrict the spaces X and Y to those for which $H_*(\Omega Y, R)$ and $H^*(X, R)$ are free R-modules of finite type. Then the Lie algebra $PH_*(\Omega Y, R)$ of primitive elements is defined. For $R = \mathbb{Q}$ the Hurewicz map Φ is an isomorphism by the Milnor-Moore theorem. Dually, deg is also an isomorphism for $R = \mathbb{Q}$. For $R \neq \mathbb{Q}$ the behaviour of Φ and deg is unknown. Therefore we consider only spaces X and Y for which either Φ or deg respectively is still surjective, or else even admits a right inverse in the categories Lie_R or Divalg_R respectively.

For the twisted products $\tilde{\otimes}$ in (3) we show

(7) Theorem. (A) If Φ_Y admits a right inverse we have an isomorphism in Lie_M

$$L(Y, R) \cong PH_*(\Omega Y, R) \tilde{\otimes} M_R$$

(B) If $\deg_X$ admits a right inverse we have an isomorphism in Divalg_M

$$M(X, R) \cong M_R \tilde{\otimes} H^*(X, R).$$

Clearly (A) corresponds to the Milnor-Moore theorem for $R = \mathbb{Q}$. For $R \neq \mathbb{Q}$ the Hilton-Milnor theorem, as well as the results of G. J. Porter on homotopy groups of a fat wedge of spheres, are further illustrations of (A).

Next we study the R-localization of the group $[\Sigma X, Y] = [X, \Omega Y]$, which we assume to be nilpotent. Our results are also applicable to the study of the group $[X, G]$ where G is a topological group. We define a bifunctor

$$\exp_M : \text{Divalg}_M \times \text{Lie}_M \to \text{Category of groups}$$

which is essentially the exponential group on a Lie algebra. Furthermore we obtain a natural homomorphism

$$\rho : \exp_M(M(X, R), L(Y, R)) \to [\Sigma X, Y]_R$$

and we prove

(8) Theorem. If Φ_Y or $\deg_X$ is surjective, the homomorphism ρ is an isomorphism.

If Φ_Y or $\deg_X$ even admits a right inverse, we can replace the coefficients M by R. In fact, since we have isomorphisms

$$\exp_M(M \tilde{\otimes} A, L) = \exp_R(A, L)$$

$$\exp_M(B, K \tilde{\otimes} M) = \exp_R(B, K)$$

we obtain from (8) and (7)

(9) **Theorem.** (A) _If $\deg_X$ admits a right inverse, we have an isomorphism_

$$\exp_R(H^*(X, R), \pi_*(\Omega Y) \otimes R) \cong [\Sigma X, Y]_R .$$

(B) _If Φ_Y admits a right inverse, we have an isomorphism_

$$\exp_R(M(X, R), PH_*(\Omega Y, R)) \cong [\Sigma X, Y]_R .$$

(C) _If $\deg_X$ and Φ_Y both admit right inverses, the group $[\Sigma X, Y]_R$ depends only on the cohomology algebra $H^*(X, R)$ and on the homology Lie algebra $PH_*(\Omega Y, R)$._

Clearly for $R = \mathbb{Q}$ the propositions (A), (B) and (C) coincide. For $R = \mathbb{Q}$ theorem (9) is equivalent to

(10) **Theorem.** _There is a natural isomorphism of rational nilpotent groups_

$$[\Sigma X, Y]_{\mathbb{Q}} \cong \exp \operatorname{Hom}(H_*(X, \mathbb{Q}), \pi_*(\Omega Y) \otimes \mathbb{Q}).$$

Here the $\mathbb{Q}$-vector space of degree zero homomorphisms

$$\operatorname{Hom}(H_*(X, \mathbb{Q}), \pi_*(\Omega Y) \otimes \mathbb{Q})$$

has in a natural way the structure of a nilpotent rational Lie algebra and exp denotes the group structure on this rational vector space given by the Baker-Campbell-Hausdorff formula. With a certain amount of work, formula (10) can also be derived from the rational homotopy theories of Quillen or Sullivan. However, the formula does not appear in the literature. We give a different type of proof which is based only on the old result of Serre that the rational n-sphere $S^n_{\mathbb{Q}}$ is an Eilenberg-MacLane space if n is odd.

As a generalization of (10) we obtain, for example, from (9) (A) and (3.9) in chapter V

(11) Theorem. Let $H_*(X, R)$ be a finitely generated free R-module. Let X be connected and let G be a connected topological group. Then there is a natural isomorphism

$$[X, G]_R = \exp_R(H^*(X, R), \pi_*(G) \otimes R)$$

of R-local nilpotent groups if we assume that R contains 1/p for all primes p with

$$p < \tfrac{1}{2}(\dim_R(X) - C_R(X) + 3) :$$

$\dim_R X$ denotes the top dimension n with $H^n(X, R) \neq 0$, and $C_R(X)$ is the smallest dimension $n \geq 1$ with $H^n(X, R) \neq 0$. The reason for the inequality in the theorem is that the homotopy group $\pi_m(S^n)$ of a sphere S^n has no p-torsion for $p < \frac{1}{2}(m - n + 3)$.

These results make it already sufficiently apparent that indeed we can exploit rational homotopy theory in the non-rational case. Unfortunately, a difficulty in the way of such an approach is that the methods of proof in rational homotopy theory are not at all available in the non-rational situation. Thus an entirely new approach is necessary.

INTRODUCTION TO PART A

We obtain the results of part B by an extensive and systematic study of the algebraic properties of the classical homotopy operations

composition of maps $\circ$
smash products $\#$, $\underline{\#}$
Whitehead product $[\ ,\ ]$
James-Hopf invariants γ_n
addition $+$

It is much easier and of more general interest to consider these operations in their generalized form, namely

$$\circ : [\Sigma A, \Sigma B] \times [\Sigma B, Z] \to [\Sigma A, Z]$$

$$\#, \underline{\#} : [\Sigma X, \Sigma A] \times [\Sigma Y, \Sigma B] \to [\Sigma X \wedge Y, \Sigma A \wedge B]$$

$$[\ ,\ ] : [\Sigma A, Z] \times [\Sigma B, Z] \to [\Sigma A \wedge B, Z]$$

$$\gamma_n : [\Sigma A, \Sigma B] \to [\Sigma A, \Sigma B^{\wedge n}]$$

$$+ : [\Sigma A, Z] \times [\Sigma A, Z] \to [\Sigma A, Z].$$

Many formulas relating to these operations are scattered through the literature. In the beginnings of homotopy theory the operations were only considered on homotopy groups $\pi_n(Y) = [S^n, Y]$. It took some time before the significance of the generalized operations became evident. The Whitehead product was invented by J. H. C. Whitehead in 1941. Arkowitz and Barratt obtained its generalization around 1960. In 1955 James gave his wonderful combinatorial definition of the higher-order Hopf invariants γ_n that is fundamental to our work. The nature of the higher invariants $(n > 2)$ remained unclear. They later were more systematically studied by Boardman and Steer (1967). However, their point of view is too stable-minded for our purposes, since they only consider the suspended invariants

$$\lambda_n(\alpha) = \Sigma^{n-1} \gamma_n(\alpha) .$$

Moreover, they still use the left distributivity law for expanding the composite $(\xi + \eta)\alpha$ in terms of the Hilton-Hopf invariants as it was presented by Hilton in 1955. In this book we exhibit a more agreeable left distributivity law in terms of the James-Hopf invariants, namely

$$(*) \qquad \xi \circ \alpha + \eta \circ \alpha = (\xi + \eta) \circ \alpha + \sum_{n \geq 2} c_n(\xi, \eta) \circ \gamma_n(\alpha) .$$

(We use this formula to prove explicitly the folklore result that the James-Hopf invariants determine the Hilton-Hopf invariants.) A further major result of this book is the expansion formula for the Whitehead product $[\xi \circ \alpha, \eta \circ \beta]$ of composition elements $\xi \circ \alpha$ and $\eta \circ \beta$. This formula again uses the James-Hopf invariants and is of the form

$$(**) \qquad [\xi \circ \alpha, \eta \circ \beta] = \sum_{n \geq 1} \sum_{m \geq 1} R_{m,n}(\xi, \eta) \circ (\gamma_m(\alpha) \mathbin{\underline{\#}} \gamma_n(\beta)) .$$

A special case of this formula was already found by Barcus-Barratt in 1958. The terms $c_n(\xi, \eta)$ and $R_{m,n}(\xi, \eta)$ are sums of iterated Whitehead products in ξ and η. We construct these terms explicitly. These two expansion formulas are basic to the development of our theory. Their proof makes use of classical commutator calculus in nilpotent group theory and Lie algebra theory. Chapter I therefore is purely algebraic. Various results of chapter I, while motivated by homotopy theory, seem to be new. They also may be of interest in combinatorial group theory and Lie algebra theory.

One of our crucial observations is that the above expansion formulas (*) and (**) are in fact closely connected with the following two formulas for the exponential function

$$e^x = \sum_{n \geq 0} x^n/n!$$

in a free tensor algebra. The Baker-Campbell-Hausdorff formula presents an infinite sum $\Phi(x, y)$ of <u>rational</u> Lie elements with the property

$$e^x e^y = e^{\Phi(x, y)} .$$

Evaluated in a rational nilpotent Lie algebra, the sum $\Phi(x, y) = x\ y$ becomes finite and gives us a group multiplication on L. This group we denoted by exp(L). We prove that there exist integral Lie elements $c_n(x, y)$ and $R_{m,n}(x, y)$ such that for $x, y \in L$ the group multiplication in exp(L) satisfies the equations

$$(*)' \qquad x\ y = (x + y) \prod_{n\geq 2} c_n(x, y)^{\frac{1}{n!}}$$

$$(**)' \qquad x^{-1}y^{-1}xy = \prod_{n\geq 1} \prod_{m\geq 1} R_{m,n}(x, y)^{\frac{1}{m!}\frac{1}{n!}}$$

These equations are essentially special cases of the expansion formulas (*) and (**) respectively. The characterization of the terms $c_n(x, y)$ and $R_{m,n}(x, y)$ as a sum of iterated Lie brackets in x and y is about the same as the one of the corresponding terms in (*) and (**). The factors of the form 1/m! in the formulas (*)' and (**)' correspond to the James-Hopf invariant $\gamma_m(\ )$. In fact, this book will make it plausible to the reader that the James-Hopf invariants can be regarded as divided power operations. For this reason we believe that the theory of Malcev and Lazard on the exponential correspondence between nilpotent rational Lie algebras and nilpotent rational groups allows a still further generalization by using divided power operations. We will not take the algebra that far. However, a step in this direction is our construction of the group $\exp_{\mathbb{Q}}(A, L)$ where A is a (say finitely generated) graded commutative rational algebra and L is a graded rational Lie algebra. The group $\exp_{\mathbb{Q}}(A, L)$ is generated by all pairs (x, α) with $x \in A^n$, $\alpha \in L_n$ and $n \geq 1$. The principal relations are

$$(*)'' \qquad (x, \alpha)(x, \beta) = (x, \alpha + \beta) \prod_{n\geq 2} \left(\frac{x^n}{n!}, c_n(\alpha, \beta)\right)$$

$$(**)'' \qquad (x, \alpha)^{-1}(y, \beta)^{-1}(x, \alpha)(y, \beta) = \prod_{n\geq 1} \prod_{m\geq 1} \left(\frac{x^m}{m!}\frac{y^n}{n!}, R_{m,n}(\alpha, \beta)\right)$$

which correspond to (*) and (**) above. The algebra structure on A induces a coalgebra structure Δ on the dual space C with $C_n = \mathrm{Hom}(A^n, \mathbb{Q})$. Moreover the $\mathbb{Q}$-vector space Hom(C, L) of degree zero homomorphisms is a Lie algebra by the bracket

$$[f, g] = [\ ,\ ] \circ (f \otimes g) \circ \Delta : C \to C \otimes C \to L \otimes L \to L$$

For this Lie algebra we obtain a canonical isomorphism of groups

(12) Theorem. $\exp_{\mathbb{Q}}(A, L) = \exp \mathrm{Hom}(C, L)$.

In the definition of $\exp_{\mathbb{Q}}(A, L)$ we can replace $\mathbb{Q}$ by a ring R and the terms $x^n/n!$ by divided power operations $\gamma_n(x)$ on a commutative algebra A over R. In this way we get the group $\exp_R(A, L)$ which generalizes the exponential group in (12). In fact the homotopy theoretic analogue is much more general and gives us a group by using the relations (*) and (**) instead of (*)'' and (**)''. As we show in §5 of chapter II this group leads to a presentation of the group $[\Sigma X, Z]$ if we assume that

$$\Sigma X \simeq \bigvee_{i \in J} \Sigma Y_i$$

decomposes as a one-point union of suspended co-H-spaces Y_i. Theorem (8) above is essentially a particular case of this general result.

Working with algebras A with divided powers is more complicated than working with rational algebras. For example, assume that all powers x^n for $x \in A$ are trivial. Then in a rational algebra all divided powers $\gamma_n(x) = x^n/n!$ are clearly trivial. This is not the case if A is an algebra over a subring R of $\mathbb{Q}$. Here the divided powers have the property that for a prime power $p^\nu = n$ with $1/p \notin R$ the divided power $\gamma_n(x)$ may still be a nontrivial element of order p. This actually happens in homotopy theory. The James-Hopf invariants γ_n on homotopy groups of spheres are essentially divided power operations on an algebra with trivial powers. We present elements α in the homotopy groups of spheres for which $\gamma_p(\alpha)$ is non-zero and thus is an element of order p, see chapter IV. This peculiarity of the James-Hopf invariants, as we show, corresponds to an interesting feature of the terms $c_n(x, y)$ in (*)' above, namely that modulo the prime p we have a congruence

(13) Theorem. $c_p(x, y) \equiv x^{\otimes p} + y^{\otimes p} - (x + y)^{\otimes p} \bmod p$.

The righthand side is classically known to be a Lie element modulo p. In a similar way, we prove an old result of Zassenhaus, namely that also

$$x_1^{\otimes p^\nu} + \ldots + x_k^{\otimes p^\nu} - (x_1 + \ldots + x_k)^{\otimes p^\nu}$$

is modulo p a Lie element, and we obtain a formula for it, see (I. 4. 9). We will derive this formula from expansion formulas in homotopy theory.

I would like to acknowledge the support of the Sonderforschungsbereich 40 "Theoretische Mathematik" towards the completion of this book. Furthermore, I am grateful to the publishers for their helpful cooperation.

H. J. Baues

PART A: HOMOTOPY OPERATIONS, NILPOTENT GROUP THEORY AND NILPOTENT LIE ALGEBRA THEORY

I. COMMUTATOR CALCULUS

Commutator calculus is a branch of group theory dealing with nilpotent groups, Lie algebras, the exponential function and the Baker-Campbell-Hausdorff formula, [12, 30]. One purpose of this book is to exhibit a close connection of commutator calculus with classical homotopy theory.

In this chapter we describe some properties of the exponential function on a rational tensor algebra. So we give explicit formulas for the Zassenhaus terms and for an exponential commutator. The proof of these formulas will be given via homotopy theory in chapter II, §4. Moreover, we obtain a new presentation of the exponential group on a Lie algebra of homomorphisms. In chapter II, §5, we exhibit the homotopy theoretic analogue of this presentation, see also Chapter VI. In §4 we characterize the Zassenhaus terms $c_{p^n}(x, y)$ modulo a prime p, for example we show

$$c_p(x, y) \equiv x^{\otimes p} + y^{\otimes p} - (x + y)^{\otimes p} \mod p.$$

Both sides of this equation are well known terms in classical commutator calculus [30], however, that they are equal $\bmod p$ seems to be a new result. These equations will be of importance in the proof (VII § 3) that there exists the M-extension of a Lie algebra where M is a module of homotopy coefficients.

§ 1. The exponential function and the Zassenhaus formula

Let V be a $\mathbb{Q}$ vector space and let

$$(1.1) \quad T(V) = \bigoplus_{n \geq 0} V^{\otimes n} \subset \hat{T}(V) = \prod_{n \geq 0} V^{\otimes n}$$

be the tensoralgebra on V. If $V = \{ V_m | m \in \mathbb{Z} \}$ is graded, then $V^{\otimes n}$ and $T(V)$ are graded $\mathbb{Q}$ vector spaces. The Lie bracket is defined by

(1.2) $[x, y] = xy - (-1)^{|x||y|} yx$

where x, y are homogeneous terms of T(V) of degree $|x|$, $|y|$. If V is not graded, we regard V as concentrated in degree 0, so that in this case we obtain the ordinary Lie bracket

(1.3) $[x, y] = xy - yx.$

The free Lie algebra L(V) is the sub Lie algebra generated by V in T(V). $x \in T(V)$ is a Lie element if $x \in L(V)$. We say x has weight n if $x \in V^{\otimes n} \subset T(V)$.

For the remainder of this section let V be a non graded $\mathbb{Q}$-vector space. Thus the Lie bracket is given by (1.3). $L_{\mathbb{Z}}(x, y)$ denotes the free Lie algebra over $\mathbb{Z}$ generated by the elements x and y. We call elements of $L_{\mathbb{Z}}(x, y)$ integral Lie elements (in x and y). They are integral linear combinations of iterated Lie brackets with all factors being x or y. Clearly for $x, y \in V$ we have a canonical map $L_{\mathbb{Z}}(x, y) \to L(V)$.

For $x \in V$ the exponential function e^x is defined by the infinite sum

(1.4) $e^x = \sum_{n=0}^{\infty} x^n/n! \in \hat{T}(V).$

The Baker-Campbell-Hausdorff formula states that

(1.5) $e^x e^y = e^{\phi(x, y)} \qquad (x, y \in V)$

where

(1.6) $$\phi(x, y) = x + y + \tfrac{1}{2}[x, y] + \tfrac{1}{12}[[x, y],y] + \tfrac{1}{12}[[y, x], x] + \dots$$

is an (infinite) sum of homogeneous Lie elements, see [26], [30]. (1.6) gives the first terms of $\phi(x, y)$ up to and including terms of weight 3. Similar to the Zassenhaus formula (see p. 372 in [30]) we obtain:

I

(1.7) **Proposition.** <u>There exist integral Lie elements</u> $c_n(x, y)$ <u>of weight</u> n <u>such that</u>

$$e^x e^y = e^{x+y} \prod_{n\geq 2} e^{c_n(x,y)/n!}$$

The first terms are

$$c_2(x, y) = [x, y]$$

$$c_3(x, y) = 2[[x, y], y] + [[x, y], x]$$

$$c_4(x, y) = [c_3(x, y), x] + 3[[x, y, y, y]$$
$$+ [[x, y, x, y] + [[x, y], [x, y]]$$

We use the notation

$$[[x_1, \ldots, x_n] = [[\ldots[x_1, x_2], \ldots, x_{n-1}], x_n].$$

Clearly the last summand of $c_4(x, y)$ is trivial over $\mathbb{Q}$, compare remark (1.18).

We now exhibit a method of computing the terms $c_n(x, y)$. The natural numbers are ordered by $1 < 2 < 3 < \ldots$, we say a function α is monotone if $x \leq y$ implies $\alpha x \leq \alpha y$.

(1.8) **Definition.** Let $P(\mathbb{N})$ be the set of all subsets of $\mathbb{N} = \{1,2,3,\ldots\}$. For $a \in P(\mathbb{N})$ we write $a = \{a_1 < a_2 < \ldots < a_{\#a}\}$ where $\#a$ is the number of elements of a. We say a total ordering $<$ on $P(\mathbb{N})$ is <u>admissible</u> if for $a, b \in P(\mathbb{N})$

(i) $a < b \Rightarrow \#a \leq \#b$,

(ii) for any monotone injective function $\alpha : \mathbb{N} \to \mathbb{N}$ the induced function $\alpha : P(\mathbb{N}) \to P(\mathbb{N})$ is monotone,

(iii) for $\{1\}, \{2\} \in P(\mathbb{N})$ let $\{1\} < \{2\}$.

We denote with $\overset{<}{\sum}_{a\subset\bar{n}} r_a$ the sum of elements r_a taken in the ordering $<$ over all indices $a \in P(\mathbb{N})$ with $a \subset \bar{n} = \{1, \ldots, n\}$.

Remark. (ii) and (iii) imply that the function $\beta : \mathbb{N} \to P(\mathbb{N})$ with $\beta(i) = \{i\}$ is monotone.

Examples of admissible orderings on $P(\mathbb{N})$ are the lexicographical orderings from the left or from the right.

Example. The lexicographical ordering from the left on $P(\mathbb{N})$ is defined by

$$a < b \iff \#a < \#b \text{ or in case } \#a = \#b \text{ with } a_i = b_i \text{ for } i < j \text{ and } a_j \neq b_j \text{ then } a_j < b_j.$$

(1.9) Definition. Let $F(M)$ be the free non associative algebraic object with one binary operation $[\ ,\ ]$ generated by the set M. $F(M)$ is the set of iterated brackets in letters $x_1, \ldots, x_k \in M$, $k \geq 1$. Let $|x|$ be the length of the bracket $x \in F(M)$, that is, the total number of factors in x. Let $FG(M)$ be the free group generated by M and let $L_{\mathbb{Z}}(M)$ be the free Lie algebra over $\mathbb{Z}$ generated by M. We have canonical functions (which we suppress from the notation)

$$L_{\mathbb{Z}}(M) \leftarrow F(M) \rightarrow FG(M)$$

mapping a bracket to a commutator and to a Lie bracket respectively. We write the multiplication in $FG(M)$ additively, so that $[x, y] = -x - y + x + y$ is the commutator.

(1.9)' Notation. If $D(z_1, \ldots, z_n)$ is a subset of $F(z_1, \ldots, z_n)$ then $D(v_1, \ldots, v_n)$ denotes the corresponding subset of $F(v_1, \ldots, v_n)$ obtained by the bijection $z_i \mapsto v_i$, $i = 1, 2, \ldots, n$.

Let $G = FG(x_1, x_2, \ldots, y_1, y_2, \ldots)/\sim$ be the group given by the relations $[x, y] \sim 0$ if $\underline{x} \cap \underline{y} \neq \emptyset$. The set $\underline{x} \subset \mathbb{N}$ for $x \in FG(x_1, x_2, \ldots, y_1, y_2, \ldots)$ is the set of all indices of letters in x written as a word in reduced form. For the group G we derive the following lemma which is crucial for the computation of the Zassenhaus terms.

(1.10) Lemma. For any admissible ordering $<$ on $P(\mathbb{N})$ there exist subsets

$$D_n = D(x_1, \ldots, x_n, y_1, \ldots, y_n) \subset F(x_1, \ldots, x_n, y_1, \ldots, y_n)$$

I

of elements of length n, $n \geq 1$, such that in G we have the equation

$$x_1 + x_2 + \ldots + x_n + y_1 + y_2 + \ldots + y_n =$$

$$(x_1 + y_1) + \ldots + (x_n + y_n) + \sum^{<}_{a \subset \bar{n}} d_a \ .$$

For $a = \{a_1 < \ldots < a_{\#a}\} \subset \bar{n}$ the element $d_a \in G$ is the sum of all iterated commutators $d \in D(x_{a_1}, \ldots, x_{a_{\#a}}, y_{a_1}, \ldots, y_{a_{\#a}})$ given by $D_{\#a}$, see (1.9)'. The sum d_a can be taken in arbitrary order.

If any index appears twice in $d \in D_n$ by our assumption on G the element d is trivial in G. Therefore D_n in (1.10) can be chosen such that we have a function

(1.11) $\tau : D_n \to S_n$

where S_n is the permutation group of $\bar{n}$. $\tau(d)$ is the permutation mapping $i \in \bar{n}$ to the index of the i-th factor (from the left) of d. By forgetting indices we have a mapping

(1.12) $\Phi : F(x_1, x_2, \ldots, y_1, y_2, \ldots) \to F(x, y)$.

In §4 of chapter II we will prove:

(1.13) Theorem. For any choice of an admissible ordering $<$ on $P(\mathbb{N})$ and of subsets $D_n \subset F(x_1, x_2, \ldots, y_1, y_2, \ldots)$ as obtained in (1.10) the elements

$$c_n(x, y) = \sum_{d \in D_n} \Phi(d)$$

satisfy the equation in (1.7). Clearly in this sum $\Phi(d)$ denotes an integral Lie element in $L_{\mathbb{Z}}(x, y)$.

We shall see that (1.13) is a special case of a homotopy theoretic result on higher order Hopf invariants, see II (2.8).

The following proof of lemma (1.10) gives an inductive construction for the sets D_n. This yields a description of all terms $c_n(x, y)$ by (1.13).

In any group we have the Witt-Hall identity (see page 290 of [30])

(1.14) $[x, y+z] = [x, z] + [x, y] + [[x, y], z]$.

Now let $\overline{G} = FG(x, y_1, \ldots, y_n)/\sim$ be the group given by the relations $[y, y'] \sim 0$ if $\underline{y} \cap \underline{y}' \neq \emptyset$, $\underline{y} \subset \mathbb{N}$ denotes the set of all indices i of letters y_i in y written as a word in reduced form. We derive from (1.14) by induction (take $y = y_1$ and $z = y_2 + \ldots + y_n$):

(1.15) **Lemma.** In $\overline{G}$ we have the equation

$$[x, y_1 + \ldots + y_n] = Y_n + Y_{n-1} + \ldots + Y_1$$

where

$$Y_j = \sum_{a \subset \overline{n},\ j=\mathrm{Min}(a)} [[x, y_{a_1}, \ldots, y_{a_{\#a}}].$$

The sum Y_j can be taken in arbitrary order. $\mathrm{Min}(a) = a_1$ denotes the smallest element of a.

Proof of lemma (1.10). We set $D_1 = \emptyset$. Assume now we have found sets D_k, $k \leq n$, such that the equation in (1.10) is valid for these D_k. Then we obtain D_{n+1} as follows. We consider the group G with relations as in (1.10). Comparing with (1.15) we observe first that

$$x_{n+1} + (y_1 + \ldots + y_n) = (y_1 + \ldots + y_n) + x_{n+1} + U$$

where $U = \sum_{b \subset \overline{n},\ b \neq \emptyset} z_b$ and $z_b = [[x_{n+1}, y_{b_1}, \ldots, y_{b_{\#b}}]$. By definition of G all summands of U commute in G. Now we know from the inductive assumption that

$$\begin{aligned} x_1 + \ldots + x_{n+1} + y_1 + \ldots + y_{n+1} &= x_1 + \ldots + x_n + y_1 + \ldots + y_n + x_{n+1} + U + y_{n+1} \\ &= (x_1 + y_1) + \ldots + (x_n + y_n) + \\ &\quad + \sum_{a \subset \overline{n}}^{<} d_a^n + x_{n+1} + U + y_{n+1} \end{aligned}$$

with $d_a^n = d_a$ as in (1.10). Here U and y_{n+1} commute since all summands of U have the factor x_{n+1}.

With $z_\emptyset = x_{n+1} = y_{n+1}$ a collection process yields

I

$$(*)\qquad \overset{<}{\sum_{a\subset\bar{n}}} d^n_a + \sum_{b\subset\bar{n}} z_b = z_\emptyset + \overset{<}{\sum_{a\subset\overline{n+1}}} d^{n+1}_a$$

and thus the sets D_{n+1}. Clearly all elements z_b commute and they do not appear twice in a commutator created by the collecting process. We have to collect the sum at the left side of (*) in such a way that the index sets appear in the right order $<$. Therefore we bring z_b having index set $b \cup \{n+1\}$ to its place $b \cup \{n+1\}$. This creates the commutators

$$[d^n_{a^1}, z_b] \text{ for } a^1 \subset \bar{n} \text{ and } a^1 > b \cup \{n+1\}.$$

This commutator (created at place a^1) has to be brought to its place $a^1 \cup b \cup \{n+1\}$. This creates the commutators

$$[[d^n_{a^1}, z_b], d^n_{a^2}] \text{ for } a^1 < a^2 < a^1 \cup b \cup \{n+1\}$$

since we know $a^1 \leq a^1 \cup b \cup \{n+1\}$. Inductively we obtain

$$d^{n+1}_a = d^n_a \text{ if } a \subset \bar{n}$$

and if $a = \tilde{a} \cup \{n+1\}$, $\emptyset \neq \tilde{a} \subset \bar{n}$, we have

$$(**)\qquad d^{n+1}_a = z_{\tilde{a}} + \sum [[d^n_{a^1}, z_b, d^n_{a^2}, \ldots, d^n_{a^k}]$$

where we sum over all $b \subset \tilde{a}$ and partitions $(a^1, \ldots, a^k)$ of $\tilde{a} - b$, $k \geq 1$, with

$$b \cup \{n+1\} < a^1 < a^2 < \ldots < a^k \text{ and } a^s < a^1 \cup \ldots \cup a^{s-1} \cup b \cup \{n+1\}$$

for $s = 2, \ldots, k$. By definition of the relations in G all index sets $a^1, b, a^2, \ldots, a^k$ are disjoint. We also know that the iterated commutator in (**) is multilinear in the summands of $z_\emptyset$ and $d^n_{a^i}$ $(i = 1, \ldots, k)$, compare (1.14).

If we consider the set of summands of $d^{n+1}_{\overline{n+1}}$ in (**) we obtain:

(1.16) Definition. Let $<$ be an admissible ordering on $P(\mathbb{N})$. We define inductively over n subsets

$$D_n = D(x_1, \ldots, x_n, y_1, \ldots, y_n) \subset F(x_1, \ldots, x_n, y_1, \ldots, y_n)$$

Let $D_1 = \emptyset$ and assume D_k is defined for $k \le n$, $n \ge 1$. Then D_{n+1} is the set containing $[[x_{n+1}, y_1, \ldots, y_n]$ and all brackets

$$[[d(a^1), \delta(b), d(a^2), \ldots, d(a^k)]$$

where

(1) $b \subset \bar{n}$ and $(a^1, \ldots, a^k)$ is a partition of $\bar{n} - b$, $k \ge 1$, with $b \cup \{n+1\} < a^1 < \ldots < a^k$ and $a^s < a^1 \cup a^2 \cup \ldots \cup a^{s-1} \cup b \cup \{n+1\}$ for $s = 2, \ldots, k$.

(2) for $a \in \{a^2, \ldots, a^k\}$ with $a = \{a_1 < \ldots < a_{\#a}\}$ the element $d(a)$ is any element in the set

$$D(x_{a_1}, \ldots, x_{a_{\#a}}, y_{a_1}, \ldots, y_{a_{\#a}}).$$

These sets are already defined by the inductive assumption, compare (1.9)'.

(3) for $b = \emptyset$ empty $\delta(\emptyset)$ is any of the two elements x_{n+1} and y_{n+1} and for $b = \{b_1 < \ldots < b_{\#b}\}$ non empty $\delta(b) = [[x_{n+1}, y_{b_1}, \ldots, y_{b_{\#b}}]$.

For the proof of (1.10) we have still to check that d_a^{n+1} as obtained in (**) for $a \subset \bar{n}$, $a \ne \bar{n}$, has the property described in (1.10), that is, d_a^{n+1} is given via convention (1.9)' by $D_{1+\#a}$. However, this is a consequence of the inductive definition (1.16) and of the assumption that $<$ is an admissible ordering on $P(\mathbb{N})$. This completes the proof of (1.10). //

(1.17) **Example.** Assume $<$ is the lexicographical ordering from the left on $P(\mathbb{N})$. Then we obtain the first examples of sets D_n as follows:

$$D_2 = \{[x_2, y_1]\},$$

$$D_3 = \{[[x_3, y_1, y_2]\} \cup \{[[x_2, y_1, \delta] \mid \delta \in \{x_3, y_3\}\}$$

$$D_4 = \{[[x_4, y_1, y_2, y_3]\} \cup$$
$$\{[d, \delta] \mid \delta \in \{x_4, y_4\} \text{ and } d \in D_3\} \cup$$
$$\{[[x_3, y_2], [x_4, y_1]]\}.$$

If we apply formula (1.13) we obtain the formulas for $c_n(x, y)$ in (1.7)

I

for $n = 2, 3, 4$.

(1.18) **Remark.** The example D_4 shows that D_n might contain summands of $c_n(x, y)$ which are trivial over $\mathbb{Q}$, for example $[[x, y], [x, y]]$. In fact the terms of D_n have more general significance in the expansion formula of chapter II. For Whitehead products the bracket $[x, x]$ need not vanish as it does in a non graded Lie algebra over $\mathbb{Q}$.

§2. The exponential commutator

In a similar way as we constructed the Zassenhaus terms we here exhibit commutator terms for an exponential commutator.

(2.1) Proposition. There exist integral Lie elements $R_{m,n}(x, y)$ of weight $m + n$, homogeneous of length m in x and homogeneous of length n in y, such that

$$e^{-x}e^{-y}e^{x}e^{y} = \prod_{n\geq 1} \prod_{m\geq 1} e^{R_{m,n}(x,y)/(m!n!)}.$$

Since we first take the product over m and then over n we cannot expect $R_{m,n}(x, y)$ to be symmetric in n and m. The first terms including weight 4 are:

$$R_{1,1}(x, y) = [x, y]$$

$$R_{2,2}(x, y) = -[[x, [x, y]], y] + 2[[x, y], [x, y]]$$

$$R_{m,1}(x, y) = (-1)^{m-1}[x^m, y]] \qquad (m \geq 1)$$

$$R_{1,n}(x, y) = [[x, y^n] \qquad (n \geq 1).$$

(2.2) **Remark.** By problem 1 p. 372 in [30] we know that

$$e^{-x}w^{-y}e^{x} = e^{-y+Q(x,y)} \quad \text{with}$$

$$Q(x, y) = \sum_{n=1}^{\infty} [[-y, x^n]/n!$$

For $e^{-y+Q(x,y)}e^{y}$ we can apply (1. 7) to get

$$e^{-x}e^{-y}e^{x}e^{y} = \prod_{n=1}^{\infty} e^{Q_n(x,y)}$$

where $Q_1(x, y) = Q(x, y)$ and

$$Q_n(x, y) = c_n(-y + Q(x, y), y)/n!$$

for $n \geq 2$. These terms $Q_n(x, y)$ are homogeneous of length n in y, however they contain summands of arbitrarily high weight.

We now show how to compute all terms $R_{m,n}(x, y)$.

(2.3) **Definition.** A word $p = a^1 \ldots a^k$ or a tuple $p = (a^1, \ldots, a^k)$ of pairwise disjoint subsets $a^i \subset a$ with $a = a^1 \cup \ldots \cup a^k$ is a *partition* of a. We say $k = |p|$ is the length of p. Let $\mathrm{Par}(a)$ be the set of all partitions of $a \subset \mathbb{N}$ and let $\mathrm{Par}(n) = \mathrm{Par}(\{1, 2, \ldots, n\})$. Moreover let $\mathrm{PAR}(n)$ be the set of all tuples $q = (n_1, \ldots, n_k)$ with $n_i \in \mathbb{N}$ and $n_1 + \ldots + n_k = n$. There is a canonical function

$$\# : \mathrm{Par}(n) \to \mathrm{PAR}(n)$$

mapping $(a^1, \ldots, a^k)$ to $(\#a^1, \ldots, \#a^k)$. Clearly
$\#\{p \in \mathrm{Par}(n) \mid \#p = (n_1, \ldots, n_k)\} = n!/(n_1!)\ldots(n_k!)$.

(2.3)' **Notation.** If $R(n)$ is a subset of $\mathrm{Par}(n)$ we denote with $R(a)$ ($a \subset \mathbb{N}$ with $\#a = n$) the subset of $\mathrm{Par}(a)$ obtained from $R(n)$ by the bijection $i \mapsto a_i$ $(i = 1, \ldots, n)$.

Let $H = FG(P(\mathbb{N}))/\sim$ be the group given by the relations $[x, y] \sim 0$ if $\underline{x} \cap \underline{y} \neq \emptyset$. For $x \in FG(P(\mathbb{N}))$ the subset $\underline{x} \subset \mathbb{N}$ is the union of all letters x_i in the reduced word $x = x_1^{n_1} \ldots x_r^{n_r}$ with $n_i \in \mathbb{Z}$, $x_i \in P(\mathbb{N})$ $(i = 1, \ldots, r)$. For the group H we prove the following lemma from which we will derive the terms $R_{m,n}(x, y)$ in (2.1):

(2.4) **Lemma.** *For any admissible ordering* $<$ *on* $P(\mathbb{N})$ *there exist subsets* $R(n) \subset \mathrm{Par}(n)$ *such that in* H *we have the equation*

$$Y_n + Y_{n-1} + \ldots + Y_1 = \sum_{a \subset \bar{n}}^{<} r_a$$

I

where

$$Y_j = \sum_{a \subset \bar{n},\ j = \mathrm{Min}(a)} a$$

$(j = 1, \ldots, n)$ *and where*

$$r_a = \sum_{(a^1, \ldots, a^k) \in R(a),\ k \geq 1} [[a^1, a^2, \ldots, a^k]$$

is the sum of iterated commutators in H *given by* $R(\#a)$, *see* (2. 3)'. *The sums* Y_j *and* r_a *can be taken in arbitrary order.*

Lemma (1. 15) is responsible for the type of elements Y_j above, compare the proof of (3. 4) and (3. 5) in chapter II.

We use the notation

(2. 5) $[x_1, \ldots, x_n]] = [x_1, [x_2, \ldots, [x_{n-1}, x_n] \ldots]]$.

Moreover we define the brackets of length $(n + 1)$

$$[[x, y^n] = [[x, y, \ldots, y]$$

$$[x^n, y]] = [x, \ldots, x, y]]$$

and the brackets of length $n + m$

$$[x^n, y^m] = [[[x^n, y]], y^{m-1}] .$$

In §4 of chapter II we will prove:

(2. 6) Theorem. *For any choice of an admissible ordering* $<$ *on* $P(\mathbb{N})$ *and of subsets* $R(n) \subset \mathrm{Par}(n)$ *as obtained in* (2. 4) *the elements*

$$R_{m,n}(x, y) = \sum_{\substack{(a^1, \ldots, a^k) \in R(n) \\ (m_1, \ldots, m_k) \in PAR(m) \\ k \geq 1}} \frac{(-1)^{m-k} m!}{(m_1!) \ldots (m_k!)} [[[x^{m_1}, y^{\#a^1}], \ldots, [x^{m_k}, y^{\#a^k}]]$$

satisfy the equation in (2. 1).

The proof of lemma (2. 4) contains an inductive construction of the sets $R(n)$, $n \geq 1$, and thus gives a more explicit description of the commutator terms $R_{m,n}(x, y)$ by (2.6).

Proof of lemma (2.4). We set $R(1) = \mathrm{Par}(1)$. Assume now we have found sets $R(k) \subset \mathrm{Par}(k)$ $(k \leq n)$ satisfying the proposition of (2.4). Then we obtain $R(n+1)$ as follows.

For $a \in P(\mathbb{N})$ with $\#a \leq n$ let $r_a^n = r_a$ be defined as in (2.4). Since the ordering $<$ on $P(\mathbb{N})$ is admissible we can apply the monotone bijection $\alpha : \bar{n} \approx \{2, 3, \dots, n+1\}$ to the equation in (2.4). This way we get

$$\sum_{j=n+1}^{2} \sum_{y \subset \overline{n+1},\ \mathrm{Min}(y)=j} y = \sum^{<}_{a \subset \{2,\dots,n+1\}} r_a^n$$

Now, for the construction of $R(n+1)$ we set up a collection process for the left side of the following equation

$$(*) \qquad \sum_{a \subset \{2,\dots,n+1\}} r_a^n + \sum^{<}_{\substack{y \subset \overline{n+1} \\ \mathrm{Min}(y)=1}} y = \sum^{<}_{a \subset \overline{n+1}} r_a^{n+1}$$

in such a way that we obtain a correct ordering of summands. If we bring y to its place y we create the commutators $[r^n_{a^1}, y]$ for $a^1 \subset \{2,\dots,n\}$, $a^1 > y$, at place a^1. If we bring such a commutator to its place $a^1 \cup y$ (where we know $a^1 \cup y > a^1$ since $\#(a^1 \cup y) > \#a^1$) we obtain the commutators $[[r^n_{a^1}, y], r^n_{a^2}]$ for $a^1 < a^2 < a^1 \cup y$ at place a^2. Inductively we get, similarly as in the proof of (1.10), $r_a^{n+1} = r_a^n$ if $a \subset \{2,\dots,n+1\}$ and for $a \subset \{1, \dots, n+1\}$ with $1 \in a$

$$(**) \qquad r_a^{n+1} = a + \sum [[r^n_{a^1}, y, r^n_{a^2}, \dots, r^n_{a^k}]$$

where we sum over all partitions $(a^1, y, a^2, \dots, a^k)$ of a with $1 \in y < a^1 < a^2 < \dots < a^k$, $k \geq 1$, and $a^s > y \cup a^1 \cup \dots \cup a^{s-1}$ for $s = 2, \dots, k$. The set of summands of $r^{n+1}_{\overline{n+1}}$ in (**) is the set $R(n+1)$ defined as follows:

(2.7) **Definition.** Let $<$ be an admissible ordering on $P(\mathbb{N})$. We define inductively over n subsets $R(n) \subset \mathrm{Par}(n)$. Let $R(1) = \mathrm{Par}\,(1)$ and assume $R(k)$ is defined for $k \leq n$, $n \geq 1$. Then $R(n+1)$ is the set of all partitions

I

$$(r(a^1),\ y,\ r(a^2),\ \ldots,\ r(a^i)),\quad k \geq 0,$$

in Par(n + 1) with $1 \in y \subset \overline{n+1}$ and

$$\begin{cases} y < a^1 < a^2 < \ldots < a^k \\ a^s < y \cup a^1 \cup \ldots \cup a^{s-1} \text{ for } s = 2, \ldots, k \end{cases}$$

and where $r(a^i)$ is any partition in $R(a^i)$, $i = 1, \ldots, k$. These sets are already defined by the inductive assumption, compare the convention (2. 3)'.

Again as in the proof of (1. 10) we have to check that (**) for $a \neq \overline{n+1}$ is compatible with the definition of R(n) in (2. 7). This follows from property (ii) of the ordering in (1. 8). //

(2. 8) **Example.** For the lexicographical ordering from the left we derive

R(1) = {(1)}
R(2) = {(12), (2. 1)}
R(3) = {(123), (23. 1), (3. 2. 1), (2. 1. 3)}
R(4) = {(1234), (234. 1), (23. 14), (34. 12), (24. 13), (3. 2. 14), (4. 3. 12), (4. 2. 13), (34. 2. 1), (3. 2. 4. 1), (4. 3. 2. 1), (2. 1. 3. 4)}.

Here we use abbreviated notation for a partition: For example (34. 2. 1) denotes ({3, 4}, {2}, {1}).

§3. A presentation for the exponential group

A group G or a Lie algebra L is nilpotent if there exists an integer $k \geq 1$ such that an iterated bracket of any k of its elements taken in any order is zero. In a group G the bracket is the commutator and in a Lie algebra L the Lie product.

(3. 1) **Definition.** We say a nilpotent group G is a rational group if G is uniquely divisible, that is $G \to G$, $x \mapsto x^n$ is a bijection for $n \neq 0$, $n \in \mathbb{Z}$. For each nilpotent group G there is the rationalisation $G \to G_{\mathbb{Q}}$

I

where $G_{\mathbb{Q}}$ is a rational group with the universal property that any homomorphism $G \to H$ into a rational group H factors in an unique way over $G_{\mathbb{Q}}$. $G_{\mathbb{Q}}$ is also called the Malcev completion of G, see [25].

Now let L be a nilpotent Lie algebra over $\mathbb{Q}$. The Baker-Campbell-Hausdorff formula yields a group multiplication on the underlying set of L, that is, see (1. 6),

(3. 2) $\quad x \quad y = \phi(x, y)$ for $x, y \in L$.

This group, denoted as $\exp(L)$, is a rational nilpotent group. Malcev [31] (see also [12]) has shown that the construction $\exp : L \mapsto \exp(L)$ is an equivalence of categories of nilpotent rational groups and Lie algebras respectively. We deduce from (1. 7):

(3. 3) **Proposition.** _Let_ L _be a nilpotent Lie algebra over_ $\mathbb{Q}$. _Then there is an unique group multiplication on_ L _satisfying_

$$x \quad y = (x + y) \prod_{n=2}^{\infty} c_n(x, y)/n!$$

for $x, y \in L$ _and this is the multiplication given by the Baker-Campbell-Hausdorff formula._

Moreover we know from (2. 1) that the commutator in $\exp(L)$ satisfies the equation

(3. 3)' $$x^{-1}y^{-1}xy = \prod_{n \geq 1} \prod_{m \geq 1} \frac{R_{m,n}(x, y)}{n!\, m!}$$

Since L is nilpotent only a finite number of factors are non trivial.

For the special types of Lie algebras below we can deduce a new presentation of the exponential group. The following non graded Lie algebras appear naturally in homotopy theory.

(3. 4) **Definition.** Let C be a graded commutative co-algebra over $\mathbb{Q}$ (of finite type that is, C_n is a finite dimensional $\mathbb{Q}$-vector space, and let $C_0 = \mathbb{Q}$). Let π be a graded Lie algebra over $\mathbb{Q}$ with $\pi_0 = \mathbb{Q}$. Then the $\mathbb{Q}$-vector space of degree zero homomorphisms

$$\mathrm{Hom}_{\mathbb{Q}}(C, \pi)$$

I

is a non graded Lie algebra with the bracket

$$[f, g] : C \xrightarrow{\Delta} C \otimes C \xrightarrow{f \otimes g} \pi \otimes \pi \xrightarrow{[\,,\,]} \pi .$$

For example the rational homology $C = H_*(X, \mathbb{Q})$ of a space is such a coalgebra and the homotopy $\pi = \pi_*(\Omega Y) \otimes \mathbb{Q}$ of a loop space is such a Lie algebra. In chapter VI we shall prove that there is a natural isomorphism of nilpotent rational groups

(3. 5) $[X, \Omega Y]_{\mathbb{Q}} \cong \exp \mathrm{Hom}_{\mathbb{Q}}(H_*(X, \mathbb{Q}), \pi_*(\Omega Y) \otimes \mathbb{Q})$

if $H_*(X, \mathbb{Q})$ is finite dimensional, $(X, \Omega Y$ connected).

For the proof of (3. 5) we will use the following characterisation of the group multiplication on $\exp \mathrm{Hom}_{\mathbb{Q}}(C, \pi)$.

Let $A = \mathrm{Hom}(C, \mathbb{Q})$ be the dual algebra of C in (3. 1), that is, $A^n = \mathrm{Hom}(C_n, \mathbb{Q})$. Then we have a canonical function $(n \geq 1)$

(3. 6) $A^n \times \pi_n \xrightarrow{\psi} \mathrm{Hom}_{\mathbb{Q}}(C, \pi)$

$(x, \alpha) \mapsto x \otimes \alpha$

where $x \otimes \alpha$ maps t to $x(t) . \alpha$. The algebra multiplication induced by Δ on A is denoted by $\cup$. Clearly the elements $x \otimes \alpha$ generate $\mathrm{Hom}_{\mathbb{Q}}(C, \pi)$ as a $\mathbb{Q}$-vector space. The group multiplication on $\exp \mathrm{Hom}_{\mathbb{Q}}(C, \pi)$ can be characterized in terms of these generators. In the next theorem we write the group structure in a free group multiplicatively.

(3. 7) Theorem. Let C and π be as in (3. 4) and let C or π be finite dimensional. The homomorphic extension of ψ in (3. 6) yields an isomorphism of groups

$$FG\left(\bigcup_{n\geq 1} A^n \times \pi_n\right)/\sim \ \underset{\psi}{\cong} \ \exp \mathrm{Hom}_{\mathbb{Q}}(C, \pi).$$

The relation $\sim$ is generated by

(1) $(x, \alpha)^{-1}(y, \alpha)^{-1}(x + y, \alpha) \sim (\frac{x \cup y}{2}, [\alpha, \alpha])$

(2) $(x, \alpha)(x, \beta) \sim (x, \alpha + \beta) . \prod_{n\geq 2} (\frac{x^n}{n!}, c_n(\alpha, \beta))$

I

(3) $(x, \alpha)^{-1}(y, \beta)^{-1}(x, \alpha)(y, \beta) \sim \prod_{n\geq 1} \prod_{m\geq 1} (\frac{x^m \cup y^n}{m!\, n!}, R_{m,n}(\alpha, \beta))$

(4) $(x, r\alpha) \sim (rx, \alpha)$ for $r \in \mathbb{Q}$

where $x, y \in A$, $\alpha, \beta \in \pi$. Clearly in (1) we have $|x| = |y| = |\alpha|$ and in (ii) we have $|x| = |\alpha| = |\beta|$.

We have $(x, 0) \sim (0, \alpha) \sim 1$, so the products in (ii) and (iii) have only a finite number of factors. In particular, if degree $|x|$ is odd we know $x^2 = 0$. Thus $c_n(\alpha, \beta)$ is needed only if $|\alpha| = |\beta|$ is even. In this case, we evaluate $c_n(\alpha, \beta)$ in the graded Lie algebra π. Similar remarks apply to $R_{m,n}(\alpha, \beta)$.

It is easily seen that ψ in (3.6) satisfies the relations in (3.7) since we have

(3.8) $[x \otimes \alpha, y \otimes \beta] = (x \cup y) \otimes [\alpha, \beta]$

for the Lie bracket in (3.4). For this, it is important that C_n and $R_{m,n}$ are in fact homogeneous terms. This is the advantage of $R_{m,n}$ over Q_n in (2.2).

Theorem (3.7) can be proved along the same lines as (5.9) in chapter II.

§4. The general type of Zassenhaus terms and its characterization modulo a prime

We first generalize the Zassenhaus formula (1.7) for the case of more than two variables

(4.1) Proposition. There exist integral Lie elements $c_n(x^1, x^2, \ldots, x^k)$ of weight n such that

$$e^{x^1} e^{x^2} \ldots e^{x^k} = e^{x^1 + \ldots + x^k} \cdot \prod_{n\geq 2} e^{c_n(x^1, \ldots, x^k)/n!}$$

These terms can be computed by the following lemma which generalizes (1.10):

Let

$$G = FG(x_i^j \mid j = 1, \ldots, k, \ i \geq 1)/\sim$$

I

be the group given by the relations $[a, b] \sim 0$ if $\underline{a} \cap \underline{b} \neq \emptyset$. The set $\underline{a} \subset \mathbb{N}$ for $a \in FG(x_i^j \mid j = 1, \ldots, k, i \geq 1)$ is the set of all lower indices i of letters x_i^j in a, written as a word in reduced form.

(4.2) Lemma. For any admissible ordering $<$ on $P(\mathbb{N})$ there exist subsets

$$D_n^k \subset F(x_i^j \mid i = 1, \ldots, n, j = 1, \ldots, k)$$

of elements of length n, $n \geq 1$, such that in G we have the equation:

$$(x_1^1 + \ldots + x_n^1) + (x_1^2 + \ldots + x_n^2) + \ldots + (x_1^k + \ldots + x_n^k) =$$

$$(x_1^1 + \ldots + x_1^k) + (x_2^1 + \ldots + x_2^k) + \ldots + (x_n^1 + \ldots + x_n^k) + \sum_{a \subset \bar{n}}^{<} d_a ,$$

where for $a = \{a_1 < \ldots < a_r\} \subset \bar{n}$ the element $d_a \in G$ is the sum of all iterated commutators $d \in D_r^k \subset F(x_{a_i}^j \mid i = 1, \ldots, r, j = 1, \ldots, k)$.

As in (1.11) we have a function

(4.3) $\tau : D_n^k \to S_n$,

$\tau(d)$ is the permutation mapping $i \in \bar{n}$ to the index of the i-th factor (from the left) of d. By forgetting lower indices we have a mapping

(4.4) $\Phi : F(x_i^j \mid i = 1 \ldots n, j = 1 \ldots k) \to F(x^1, \ldots, x^k)$

More general than (1.13) we obtain:

(4.5) Theorem. For any choice of an admissible ordering $<$ on $P(\mathbb{N})$ and of subsets D_n^k as characterized in (4.2) the elements

$$c_n(x^1, \ldots, x^k) = \sum_{d \in D_n^k} \Phi(d), \quad n \geq 1,$$

satisfy the equation in (4.1). Here $\Phi(d)$ denotes an integral Lie element in $L_{\mathbb{Z}}(x^1, \ldots, x^k)$.

We can prove this result along the same lines as we prove (1.13), see II (2.8).

For a set X we denote with FM(X) and FAG(X) the free monoid and the free abelian group generated by X. For the free monoid generated by symbols x, -x with $x \in X$ we have the surjective map

$$(4.6)\qquad FM(\{x, -x \mid x \in X\}) \xrightarrow{\pi} FAG(X)$$

mapping a word $x^1 \dots x^k$ with $x^j \in \{x, -x \mid x \in X\}$ to the sum $x^1 + \dots + x^k$. On the other hand the general Zassenhaus term c_n yields a function

$$(4.7)\qquad FM(\{x, -x \mid x \in X\}) \xrightarrow{c_n} L_{\mathbb{Z}}(X)$$

mapping the word $x^1 \dots x^k$ to the integral Lie element $c_n(x^1, \dots, x^k)$. We have the following interesting property of c_n:

(4.8) **Theorem.** If $n = p^\nu$ is a power of an odd prime p then (modulo p) c_n factors over π, that is, there is a function d_n such that the diagram

$$\begin{array}{ccc} FM(\{x, -x \mid x \in X\}) & \xrightarrow{\pi} & FAG(X) \\ c_{p^\nu} \Big\downarrow & \swarrow d_{p^\nu} & \\ L_{\mathbb{Z}}(X) \otimes \mathbb{Z}/p\mathbb{Z} & & \end{array}$$

commutes.

In IV §3 we give a geometric proof of this result for $\nu = 1$.

Theorem (4.8) is also a consequence of the following explicit description of the Zassenhaus term modulo a prime: We consider the inclusion

$$L_{\mathbb{Z}}(X) \to T_{\mathbb{Z}}(X)$$

For $v \in T_{\mathbb{Z}}(X)$ we denote with

$$v^{\otimes i} = v \otimes \dots \otimes v$$

the i-fold product in $T_{\mathbb{Z}}(X)$.

(4.9) **Theorem.** Let p be an odd prime. For $x_1, \dots, x_k \in X$ we have in $T_{\mathbb{Z}}(X) \otimes \mathbb{Z}/p\mathbb{Z}$ the formula

I

$$c_{p^\nu}(x_1, \ldots, x_k) = x_1^{\otimes p^\nu} + \ldots + x_k^{\otimes p^\nu}$$
$$- (x_1 + \ldots + x_k)^{\otimes p^\nu}$$
$$- \sum_{\substack{u+w=\nu \\ u, w \geq 1}} c_{p^u}(x_1, \ldots, x_k)^{\otimes p^w}$$

The theorem implies a result of Zassenhaus, namely that

$$x_1^{\otimes p^\nu} + \ldots + x_k^{\otimes p^\nu} - (x_1 + \ldots + x_k)^{\otimes p^\nu}$$

is in fact a Lie element mod p. In formula (63 page 93 in [44]) Zassenhaus gives a description of this element in terms of brackets.

The formula in (4. 9) can be used inductively for an explicit description of the terms c_{p^ν}. For example modulo p

(4. 10) $$c_p(x, y) \equiv x^{\otimes p} + y^{\otimes p} - (x + y)^{\otimes p}$$

$$c_{p^2}(x, y) \equiv x^{\otimes p^2} + y^{\otimes p^2} - (x + y)^{\otimes p^2}$$
$$- (x^{\otimes p} + y^{\otimes p} - (x + y)^{\otimes p})^{\otimes p}$$

$$c_p(x, y, z) \equiv x^{\otimes p} + y^{\otimes p} + z^{\otimes p} - (x + y + z)^{\otimes p}$$

The formula in (4. 9) is motivated by the following calculation with homotopy groups. The notation is explained in chapter V.

Let $(M_R^{**}, \odot, \#, \gamma_n, e^r)$ be the module of spherical homotopy coefficients and let

$$L = \pi_*(\Omega \bigvee_{x \in X} \Sigma S^k) \otimes R$$

be the homotopy Lie algebra of a wedge of spheres. For the set X we have the canonical inclusion

$$X \subset L_R(X) \subset L$$

where $L_R(X)$ is the free Lie algebra generated by X (non graded, we assume k to be even). Let x, y, z be three elements of X. For $\eta \in M^{k,j} = [\Sigma S^j, \Sigma S^k]_R$ and $e \in M^{k,k}$ (the identity of ΣS^k) we have by composition $\odot$ the elements $x \odot \eta$, $y \odot \eta$, $z \odot e \in L$. For these elements we consider the bracket

$$(4.11)\quad B = [x \odot \eta + y \odot \eta,\ z \odot e]$$

in L. This bracket can be 'expanded' in two different ways as follows:

$$\begin{aligned}(4.12)\quad B &= [x \odot \eta,\ z \odot e] + [y \odot \eta,\ z \odot e] \\ &= \sum_{n\geq 1} [x^n, z]] \odot (\gamma_n(\eta) \# e) \\ &\quad + \sum_{n\geq 1} [y^n, z]] \odot (\gamma_n(\eta) \# e), \quad \text{see II (3.4)}', \\ &= \sum_{n\geq 1} ([x^n, z]] + [y^n, z]]) \odot (\gamma_n(\eta) \# e)\end{aligned}$$

On the other hand by II (2.8) and II (3.4)'

$$\begin{aligned}(4.13)\quad B &= [(x+y) \odot \eta + \sum_{m\geq 2} c_m(x, y) \odot \gamma_m(\eta),\ z \odot e] \\ &= [(x+y) \odot \eta,\ z \odot e] + \\ &\quad \sum_{m\geq 2} [c_m(x, y) \odot \gamma_m(\eta),\ z \odot e] \\ &= \sum_{n\geq 1} [(x+y)^n, z]] \odot (\gamma_n(\eta) \# e) \\ &\quad + \sum_{m\geq 2} \sum_{n\geq 1} [c_m(x, y)^n, z]] \odot (\gamma_n(\gamma_m(\eta)) \# e)\end{aligned}$$

For an odd prime p we know $\gamma_{p^u}(\gamma_{p^w}(\eta)) = \gamma_{p^{u+w}}(\eta)$. Assume now that the Hopf invariant $\gamma_{p^\nu}(\eta) \# e$ is non trivial. Then the equations above imply that modulo the prime p in $L_R(x, y, z)$:

$$(4.14)\quad [x^{p^\nu}, z]] + [y^{p^\nu}, z]] \equiv [(x+y)^{p^\nu}, z]] + \sum_{\substack{u+w=\nu \\ u\geq 0 \\ w\geq 1}} [c_{p^w}(x, y)^{p^u}, z]]$$

For $ad(x)(z) = [x, z]$ we have $ad^n(x)(z) = [x^n, z]]$. It is well known, see [15, 30], that

$$(4.15)\quad [x^{\otimes p^\nu}, z] \equiv [x^{p^\nu}, z]] \qquad (\text{mod } p).$$

Since $T(x, y) \to T(x, y, z)$, $u \to [u, z]$ is injective (4. 15) and (4. 14) imply the proposition of theorem (4. 9).

Unfortunately we only know elements η with non trivial Hopf invariants $\gamma_p(\eta) \neq e$, p an odd prime, see IV. Therefore by the consideration above we can prove (4. 9) only for $\nu = 1$.

However, if we consider the 'universal example' of the calculation in (4. 12) and (4. 13), we obtain a proof of (4. 9) via homotopy theory in general, see VI §3.

II. DISTRIBUTIVITY LAWS IN HOMOTOPY THEORY

We show that the Zassenhaus formula corresponds in homotopy theory to the left distributivity law of the composition element

$$(\alpha + \beta) \circ \gamma .$$

Furthermore, we exhibit a distributivity law for the Whitehead product

$$[\alpha \circ \alpha', \beta \circ \beta']$$

of composition elements. This formula is related to the exponential commutator

$$e^{-x}e^{-y}e^{x}e^{y} .$$

There is a long history of such formulas in homotopy theory. P. J. Hilton gave an expansion of $(\alpha + \beta) \circ \gamma$ in his classical paper [23] which for the first time expounded the importance of commutator calculus in homotopy theory, see also [24]. A special case of our expansion formula for $[\alpha\alpha', \beta\beta']$ was found by W. Barcus and M. Barratt [5].

The connection of these formulas with the Zassenhaus formula and with the exponential commutator has not been noticed. Our expansions of $(\alpha + \beta) \circ \gamma$ and $[\alpha\alpha', \beta\beta']$ are formulas in terms of James-Hopf invariants. In this way we prove that the James-Hopf invariants determine the Hilton-Hopf invariants (a statement appearing various times in the literature, [6], [13]).

As an application of the distributivity formulas we give a solution of the following problem. Assume the suspension ΣX admits a decomposition

$$\xi : \Sigma X \simeq \bigvee_{i \in J} \Sigma Y_i$$

as a one-point union of suspended co-H-spaces Y_i. Then as a set the (in general) non abelian group of homotopy classes $[\Sigma X, Z]$ is the product of the abelian groups $[\Sigma Y_i, Z]$, $i \in J$. Describe the group structure of this product set!

In §5 we show that the group structure is determined by standard homotopy operations, namely Whitehead products, geometric cup products and James-Hopf-invariants. In chapter VI we analyze the special case of this result where all Y_i are (local) spheres.

§1. Whitehead products and cup products

We recall some basic definitions. Throughout let a space be a pointed space of the homotopy type of a CW-complex. Maps and homotopies are always base point preserving. The set of homotopy classes of maps $X \to Y$ is denoted by $[X, Y]$. It contains the trivial class $0 : X \to * \in Y$.

For the product $A \times B$ of spaces we have the cofiber sequence

$$(1.1) \quad A \vee B \overset{i}{\hookrightarrow} A \times B \overset{\pi}{\to} A \wedge B = A \times B/A \vee B$$

where $A \vee B = A \times \{*\} \cup \{*\} \times B$. The n-fold products will be denoted by $A^n = A \times \ldots \times A$ and $A^{\wedge n} = A \wedge \ldots \wedge A$.

From the unit interval $I = [0, 1]$ we define the 1-sphere $S^1 = I/\{0, 1\}$ and the n-sphere $S^n = (S^1)^{\wedge n}$. We have the comultiplication

$$(1.2) \quad \mu : S^1 \to S^1 \vee S^1$$

with $\mu(t) = (2t, *)$ for $0 \le t \le \frac{1}{2}$ and $\mu(t) = (*, 2t\text{-}1)$ for $\frac{1}{2} \le t \le 1$. $\Sigma X = S^1 \wedge X$ is called the <u>suspension</u> of X and the function space $\Omega Y = \{f : S^1 \to Y \mid f(*) = *\}$ is called the <u>loop space</u> of Y. ΣX is a Co-H-space and ΩX an H-space by the induced map

$$\begin{cases} \mu = \mu \wedge X : \Sigma X \to \Sigma X \vee \Sigma X, \\ \mu = Y^{\mu} \quad : \Omega Y \times \Omega Y \to \Omega Y. \end{cases}$$

(1.3) **Definition.** A space X together with a map $\mu : X \to X \vee X$ is a <u>Co-H-space</u> when $X \to X \vee X \hookrightarrow X \times X$ is homotopic to the diagonal map. A space Y together with a map $\mu : Y \times Y \to Y$ is an H-space when $Y \vee Y \to Y \times Y \to Y$ is homotopic to the folding map. A map f is an H-map or a Co-H-map respectively if

$$\mu(f \times f) \simeq f\mu \quad \text{or} \quad (f \vee f)\mu \simeq \mu f .$$

The maps $\mu \wedge X$ and Y^{μ} induce the same group multiplication on the homotopy sets

$$(1.4) \quad [\Sigma X, Y] = [X, \Omega Y],$$

which we denote by +. Inverses in these groups are defined by means of the map $(-1) : S^1 \to S^1$ with $(-1)(t) = 1 - t$.

If X is a Co-H-space or Y an H-space, (1.4) is an abelian group.

For a topological group or an associative H-space G we have a classifying space B_G and a group isomorphism

$$[X, G] \cong [X, \Omega B_G] .$$

The group multiplication in [X, G] is induced by the multiplication on G. Because of this isomorphism the results of this paper are applicable to the groups [X, G]. Such groups were studied in [2, 3, 4, 28, 35].

The groups of homotopy classes are equipped with various well-known operations. The first one we will describe is the Whitehead product.

The cofiber sequence (1.1) induces a short exact sequence of groups

$$(1.5) \quad 0 \to [\Sigma(A \wedge B), Z] \overset{(\Sigma\pi)^*}{\to} [\Sigma(A \times B), Z] \to [\Sigma A, Z] \times [\Sigma B, Z] \to 0$$

Let p_1, p_2 be the projections of $\Sigma(A \times B)$ onto ΣA, ΣB. The Whitehead product

$$(1.6) \quad [\ ,\ ] : [\Sigma A, Z] \times [\Sigma B, Z] \to [\Sigma(A \wedge B), Z]$$

is defined by the commutator

II

$$(\Sigma\pi)^*([\alpha, \beta]) = -p_1^*\alpha - p_2^*\beta + p_1^*\alpha + p_2^*\beta$$

for $\alpha \in [\Sigma A, Z]$, $\beta \in [\Sigma B, Z]$.

The Samelson product

$$[\ ,\ ] : [A, \Omega Z] \times [B, \Omega Z] \to [A \wedge B, \Omega Z]$$

is obtained from the Whitehead product by (1.4).

If A and B are Co-H-spaces the Whitehead product is a bilinear pairing of abelian groups. If Z is an H-space the Whitehead product is trivial, that is $[\alpha, \beta] = 0$ for all α, β. The pairing is anti-commutative:

(1.7) $$[\alpha, \beta] = -(\Sigma T)^* [\beta, \alpha]$$

where $T : A \wedge B \approx B \wedge A$ exchanges A and B. If Y_1, Y_2, Y_3 are Co-H-spaces the triple Whitehead products satisfy the following Jacobi identity. Let S_n be the permutation group of $\{1, \ldots, n\}$ and let $\varepsilon : S_n \to \{1, -1\}$ be the sign homomorphism. For three permutations $\rho, \sigma, \tau \in S_3$ with $\rho 3 = 1$, $\sigma 3 = 2$, $\tau 3 = 3$ and $y_i \in [\Sigma Y_i, Z]$ we have in $[\Sigma Y_1 \wedge Y_2 \wedge Y_3, Z]$

(1.8) $$\begin{aligned} 0 = &\varepsilon_\rho T_\rho^*[[y_{\rho 1}, y_{\rho 2}], y_1] \\ &+ \varepsilon_\sigma T_\sigma^*[[y_{\sigma 1}, y_{\sigma 2}], y_2] \\ &+ \varepsilon_\tau T_\tau^*[[y_{\tau 1}, y_{\tau 2}], y_3]. \end{aligned}$$

$T_\alpha : \Sigma Y_1 \wedge Y_2 \wedge Y_3 \to \Sigma Y_{\alpha 1} \wedge Y_{\alpha 2} \wedge Y_{\alpha 3}$ is the permutation of the factors for $\alpha \in S_3$.

If we consider the case where A and B are spheres, we obtain the Whitehead or Samelson product on homotopy groups:

(1.9) $$\pi_{n+1}(X) = [\Sigma S^n, X] = [S^n, \Omega X] = \pi_n(\Omega X).$$

If X is a 1-connected space, (1.6) provides the graded homotopy group $\pi_*(\Omega X)$ with the structure of a graded Lie-algebra. However, neither $[\alpha, \alpha]$ for $\alpha \in \pi_{2n}(\Omega X)$ nor $[[\alpha, \alpha], \alpha]$ is necessarily trivial.

To fix notation we now define a general form of iterated Whitehead products. Let $F = F(z_1, \ldots, z_r)$ be the set of iterated brackets in the letters $z_1, \ldots, z_k$, see I (1. 9). The length $|c|$ of an element $c \in F$ is the number of factors in it. For a tuple $Y = (Y_1, \ldots, Y_r)$ of spaces and a tuple $\alpha = (\alpha_1, \ldots, \alpha_r)$ of elements $\alpha_i \in [\Sigma Y_i, X]$ we define the iterated Whitehead product

(1. 10) $[\alpha_1, \ldots, \alpha_r]_c \in [\Sigma\Lambda^c(Y_1, \ldots, Y_r), X], \; c \in F(z_1, \ldots, z_r),$

by induction on the length. For $c = z_i \in F$ let $\Lambda^c Y = Y_i$ and $[\alpha]_c = \alpha_i$. For $c = (a, b) \in F$ let

$$\Lambda^c Y = (\Lambda^a Y) \wedge (\Lambda^b Y) \text{ and}$$

$$[\alpha]_c = [[\alpha]_a, [\alpha]_b]$$

Clearly $\Lambda^c Y = A^{\wedge |c|}$ for $Y_1 = \ldots = Y_r = A$.

(1. 11) **Definition.** We say X has codimension $\leq N$ if all homotopy groups $\pi_n(X)$ vanish for $n \geq N$. We say X has dimension $\leq N$ if X is homotopy equivalent to a CW-complex of dimension $\leq N$.

If all Y_i are connected then $\Sigma\Lambda^c(Y)$ is $|c|$ connected. Thus for X with codimension $\leq N$ all Whitehead products $[\alpha]_c$ vanish if $|c| \geq N$. A similar argument yields a proof of a well known result of G. W. Whitehead [43].

(1. 12) **Proposition.** Let X be a connected space. If X has finite dimension or Y has finite codimension then $[X, \Omega Y]$ is a nilpotent group.

Proof. The commutator for $f, g \in [\Sigma X, Y]$ is given by

$$-f - g + f + g = [f, g] \circ (\Sigma\tilde{\Delta}_X)$$

where $\tilde{\Delta}_X : X \to X \wedge X$ is the diagonal. This follows directly from the definition (1. 6). An iterated commutator of weight n thus factors over the diagonal $\tilde{\Delta}_X : X \to X^{\wedge n}$, which is null homotopic if $n > \dim X$. //

If Y has finite codimension $\leq N$ then for the N-skeleton X^N of X we have $[X^N, \Omega Y] = [X, \Omega Y]$, see [11].

From the definition of the Whitehead product we obtain the following commutator rule in the group

$$[\Sigma(X_1 \times \ldots \times X_n), Y]$$

For $a = \{a_1 < \ldots < a_r\} \subset \bar{n} = \{1, \ldots, n\}$ let

$$p_a : X_1 \times \ldots \times X_n \to \Lambda X_a = X_{a_1} \wedge \ldots \wedge X_{a_r}$$

be the obvious projection. Then we have for $a, b \subset \bar{n}$ and $\alpha \in [\Sigma\Lambda X_a, Y]$ and $\beta \in [\Sigma\Lambda X_b, Y]$ the <u>commutator rule</u>

(1.13) $-\alpha(\Sigma p_a) - \beta(\Sigma p_b) + \alpha(\Sigma p_a) + \beta(\Sigma p_b) = [\alpha, \beta] T_{a,b}(\Sigma p_{a \cup b})$

where

$$T_{a,b} : \Sigma\Lambda X_{a\cup b} \to \Sigma\Lambda X_a \wedge \Lambda X_b$$

is defined by $T_{a,b}(t, x_{a\cup b}) = (t, x_a, x_b)$ with $x_a = (x_{a_1}, \ldots, x_{a_r})$. Clearly, if for $i \in a \cap b$, X_i is a Co-H-space then $T_{a,b} \simeq 0$, see (1.21).

Further operations we need are the cup products.

The <u>exterior cup products</u> are pairings

(1.14) $\#, \underline{\#} : [\Sigma X, \Sigma A] \times [\Sigma Y, \Sigma B] \to [\Sigma X \wedge Y, \Sigma A \wedge B]$

defined by the compositions

$$\alpha \# \beta : \Sigma X \wedge Y \xrightarrow{\alpha \wedge Y} \Sigma A \wedge Y \xrightarrow{A \wedge \beta} \Sigma A \wedge B$$

$$\alpha \underline{\#} \beta : \Sigma X \wedge Y \xrightarrow{X \wedge \beta} \Sigma X \wedge B \xrightarrow{\alpha \wedge B} \Sigma A \wedge B$$

where $\alpha \wedge Y = \alpha \wedge 1_Y$ and where $A \wedge \beta$ is the map $1_A \wedge \beta$, up to the shuffle of the suspension coordinate. These products are associative, that is,

(1.15) $(\alpha \# \beta) \# \gamma = \alpha \# (\beta \# \gamma)$

just as for $\underline{\#}$. Furthermore we have for the interchange map $T_{AB} : A \wedge B \approx B \wedge A$

(1.16) $(\Sigma T_{AB}) \circ (\alpha \# \beta) = (\beta \underline{\#} \alpha) \circ (\Sigma T_{XY})$.

The pairings are linear in the following sense:

(1.17) $\begin{cases} (\alpha + \alpha') \# \beta = \alpha \# \beta + \alpha' \# \beta \\ \alpha \underline{\#} (\beta + \beta') = \alpha \underline{\#} \beta + \alpha \underline{\#} \beta' . \end{cases}$

Thus they are bilinear if $\# = \underline{\#}$.

(1.18) **Lemma.** <u>If</u> α <u>or</u> β <u>are Co-H-maps then</u> $\alpha \# \beta = \alpha \underline{\#} \beta$.

Proof. If α or β are suspensions the proposition is trivial, if not we use Ganea's diagram in the proof of (2.7). //

(1.19) **Corollary.** <u>For compositions</u> $\alpha' \alpha$ <u>and</u> $\beta' \beta$ <u>we have</u>

$(\alpha' \circ \alpha) \# (\beta' \circ \beta) = (\alpha' \# \beta')(\alpha \# \beta)$

<u>if</u> α <u>or</u> β' <u>are Co-H-maps. The same holds for</u> $\underline{\#}$.

If A and B are Co-H-spaces and α or β is a Co-H-map then $\alpha \# \beta$ is also a Co-H-map. We might say that Co-H-maps form an ideal with respect to the cup product.

The (interior) geometric <u>cup products</u> are defined by composing with the reduced diagonal $\tilde{\Delta} : X \to X \wedge X$,

(1.20) $\cup, \underline{\cup} : [\Sigma X, A] \times [\Sigma X, B] \to [\Sigma X, \Sigma A \wedge B]$

where $\alpha \cup \beta = (\alpha \# \beta)(\Sigma \tilde{\Delta})$, similarly for $\underline{\cup}$. If X is a Co-H-space these pairings are trivial since the reduced diagonal

(1.21) $\tilde{\Delta} \simeq \pi i \mu = 0 : X \to X \vee X \to X \times X \to X \wedge X$

is null homotopic in this case. Properties of $\#$ and $\underline{\#}$ carry over to $\cup$ and $\underline{\cup}$, in particular we derive from (1.16)

(1.22) $(\alpha \cup \beta) = (\Sigma T_{B,A})\ (\beta \underline{\cup} \alpha)$.

In chapter III we study the properties of the cup product and the Whitehead product in the particular case where the image space is a sphere.

II

§2. Hopf invariants

Hopf invariants and higher order Hopf invariants are of great importance in homotopy theory and were the subject of classical studies of M. G. Barratt, I. M. James and P. J. Hilton. The suspended Hopf invariants

$$\lambda_n(\alpha) = \Sigma^{n-1}\gamma_n(\alpha)$$

were extensively analysed by J. M. Boardman and B. Steer in [13]. However, the nature of the higher invariants, for $n \geq 3$ in particular, remains unclear. In this section we show that the invariants $\gamma_n(\alpha)$ are closely related to the Zassenhaus terms. In chapters III and IV we study the Hopf invariants $\gamma_n(\alpha)$ where α is a mapping into a sphere.

For a connected space A let $J(A)$ be the infinite reduced product of James. The underlying set of $J(A)$ is the free monoid generated by $A - \{*\}$. The topology is obtained by the quotient map

$$\bigcup_{n\geq 0} A^n \xrightarrow{\pi} J(A)$$

mapping a tuple $(x_1, \ldots, x_n)$ to the word $(\pi x_1)\ldots(\pi x_n)$ where $\pi *$ denotes the empty word in $J(A)$. $J_n(A) = \pi(A^n)$ is the n-fold reduced product of A and $A = J_1(A)$ generates the monoid $J(A)$. Let $i : A \to \Omega\Sigma A$ be the adjoint of the identity on ΣA. James [27] has shown that the extension of i

$$(2.1) \quad \begin{cases} g : J(A) \to \Omega\Sigma A \\ g(x_1 \ldots x_n) = (\ldots(i(x_1) + i(x_2))\ldots + i(x_n)) \end{cases}$$

is a homotopy equivalence. g induces the isomorphism of groups

$$(2.2) \quad [\Sigma X, \Sigma A] \cong [X, \Omega\Sigma A] \cong [X, J(A)], \ \alpha \mapsto \overline{\alpha}$$

We now fix an admissible ordering $<$ on $P(\mathbb{N})$, see I (1.8). There are mappings

$$(2.3) \quad \begin{cases} g_r : J(A) \to J(A^{\wedge r}), \ (r \geq 1) \\ g_r(x_1 \ldots x_n) = \overset{<}{\prod_a} x_{a_1} \wedge \ldots \wedge x_{a_r} \end{cases}$$

where the product is taken in the fixed order over all subsets $a \subset \{1, \ldots, n\}$ with $\#a = r$. The <u>James-Hopf invariants</u> (with respect to $<$) are the functions

(2.4) $\gamma_r : [\Sigma X, \Sigma A] \to [\Sigma X, \Sigma A^{\wedge r}], \quad r \geq 1,$

induced by g_r, that is $\overline{\gamma_r(\alpha)} = (g_r)_*(\overline{\alpha})$. Clearly γ_1 is the identity by (iii) in (1.8).

Let $\overline{g} : \Sigma J(A) \to \Sigma A$ be the adjoint of g in (2.1). Then

(2.5) $\overline{g}_r = \gamma_r(\overline{g}) : \Sigma J(A) \to \Sigma A^{\wedge r}$

is the adjoint of gg_r. It is well known that the sum

(2.6) $G = \sum_{r \geq 1} j_r \overline{g}_r : \Sigma J(A) \to \bigvee_{r \geq 1} \Sigma A^{\wedge r}$

is a homotopy equivalence, where j_r is the inclusion of $\Sigma A^{\wedge r}$ into the wedge. G is the limit of the finite sums.

If α is a suspension then it is easily seen that $\gamma_r(\alpha) = 0$ for $r \geq 2$. Co-H-maps $\alpha : \Sigma X \to \Sigma A$ need not be suspensions. They can be characterized by

(2.7) **Proposition.** <u>Let</u> X <u>be finite dimensional. Then</u> $\alpha : \Sigma X \to \Sigma A$ <u>is a Co-H-map if and only if all James-Hopf invariants</u> $\gamma_r(\alpha)$ <u>are trivial for</u> $r \geq 2$.

Proof. As Ganea has shown in [18] the diagram

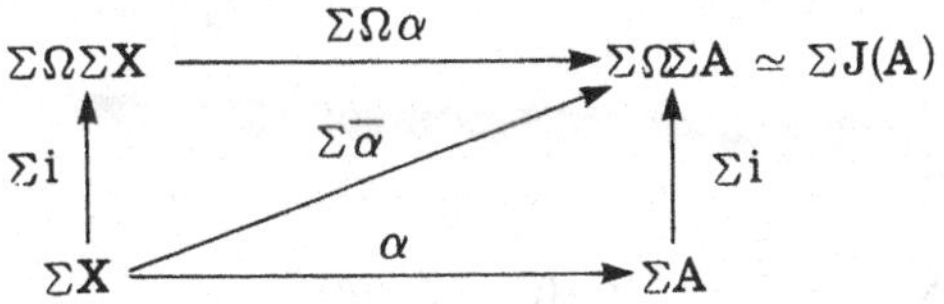

homotopy commutes iff α is a Co-H-map. Thus the result follows from (2.6). //

If α is no Co-H-map we can measure the deviation of

$$\Sigma X \xrightarrow{\alpha} \Sigma A \xrightarrow{\mu} \Sigma A \vee \Sigma A$$

from $i_1\alpha + i_2\alpha$ by use of the following expansion formula which is called the 'left distributivity law'. Here i_1, i_2 denote the two inclusions of ΣA into $\Sigma A \vee \Sigma A$, so that $\mu = i_1 + i_2$.

(2.8) Theorem. Let X be finite dimensional and let A be a Co-H-space. Then

$$i_1\alpha + i_2\alpha = (i_1+i_2)\alpha + \sum_{n\geq 2} c_n(i_1, i_2) \circ \gamma_n(\alpha)$$

where $c_n(i_1, i_2) \in [\Sigma A^{\wedge n}, \Sigma A \vee \Sigma A]$ is given by

$$c_n(i_1, i_2) = \sum_{d\in D_n} [i_1, i_2]_{\phi(d)} \circ T_{\tau(d)} .$$

The iterated Whitehead product is defined as in (1.10). D_n and τ are defined as in I, §1, and for a permutation $\sigma \in S_n$ let $T_\sigma : \Sigma A^{\wedge n} \to \Sigma A^{\wedge n}$,

$$T_\sigma(t, x_1 \wedge \ldots \wedge x_n) = (t, x_{\sigma_1} \wedge \ldots \wedge x_{\sigma_n}),$$

be the corresponding permutation of factors $x_i \in A$.

More general than (2.8) we have with the notation in I (4.2), I (4.3) and I (4.4):

(2.9) Theorem. Let X be finite dimensional and let A be a Co-H-space. Then for $\alpha \in [\Sigma X, \Sigma A]$ we have in $[\Sigma X, \Sigma A \vee \ldots \vee \Sigma A]$ the equation

$$i_1\alpha + \ldots + i_k\alpha = (i_1+\ldots+i_k)\alpha + \sum_{n\geq 2} c_n(i_1, \ldots, i_k) \circ \gamma_n(\alpha)$$

where

$$c_n(i_1, \ldots, i_k) = \sum_{d\in D_n^k} [i_1, \ldots, i_k]_{\phi(d)} \circ T_{\tau(d)}$$

Proof of (2.8). Let $R : \Sigma\Omega Y \to Y$ be the evaluation map with $R(t, \sigma) = \sigma(t)$. For the adjoint $\bar{f} : X \to \Omega Y$ of $f : \Sigma X \to Y$ we have

(1) $\quad f = R \circ (\Sigma \bar{f})$.

We consider the diagram

$$\begin{array}{ccccc} & & \bigvee_{n\geq 1} \Sigma A^{\wedge n} & & \\ & G \nearrow \simeq & & \searrow (c_n,\ n\geq 1) & \\ \Sigma X & \xrightarrow{\Sigma\bar{\alpha}} & \Sigma J(A) & \xrightarrow[i_1\bar{g} + i_2\bar{g}]{} & \Sigma A \vee \Sigma A \end{array}$$

where by use of (1) we have

(2) $\quad (i_1\bar{g} + i_2\bar{g}) \circ (\Sigma\bar{\alpha}) = i_1\alpha + i_2\alpha$

Since G is a homotopy equivalence, there exist mappings c_n making the diagram homotopy commutative. We have to show

(3) $\quad c_n = c_n(i_1, i_2), \quad c_1 = i_1 + i_2,$

as defined in (2. 8). By use of

(4) $\quad \gamma_n(\alpha) = \gamma_n(\bar{g}(\Sigma\bar{\alpha})) = \gamma_n(\bar{g}) \circ (\Sigma\bar{\alpha}) = (\bar{g}_n)_*(\Sigma\bar{\alpha})$

the equation in (2. 8) follows from (2. 6) and (3). Now we know from (2. 1) that

$$\bar{g}(\Sigma\pi) : \Sigma A^n \to \Sigma JA \to \Sigma A$$

is given by

(5) $\quad \bar{g}(\Sigma\pi) = (\Sigma p_1) + \ldots + (\Sigma p_n)$

where $p_i : A^n \to A$ is the projection to the i-th coordinate. Therefore we obtain

(6) $\quad (i_1\bar{g} + i_2\bar{g})\Sigma\pi = x_1 + \ldots + x_n + y_1 + \ldots + y_n$

where $x_i = i_1(\Sigma p_i)$ and $y_i = i_2(\Sigma p_i)$.

On the other hand we know from the definition of g_r and G in (2. 6) that

(7) $\quad G \circ (\Sigma\pi) = \sum_{r=1}^{n} \sum_{\substack{a \subset \bar{n} \\ \#a = r}}^{<} j_r \circ (\Sigma p_a)$

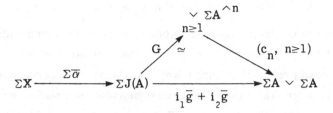

where $p_a : A^n \to A^{\wedge r}$ is the projection with $p_a(x) = x_{a_1} \wedge \ldots \wedge x_{a_r}$. The sum is taken in the fixed order on $P(\mathbb{N})$. By use of the commutator rule in $[\Sigma A^n, \Sigma A \vee \Sigma A]$ we can now 'collect' the summands of (6) (creating Whitehead products) in such a way that the newly ordered sum obviously factors over (7). Since $\Sigma\pi$ is monomorphic (see (1.5)) we have therefore computed the terms c_n. The collecting is done in lemma I (1.10). Thus (2.8) is proved. //

Hilton in [23] obtained another expansion of $(i_1 + i_2)\alpha$.

For finite dimensional X and a connected Co-H-space A the Hilton-Hopf invariants

$$(2.10) \quad h_c : [\Sigma X, \Sigma A] \to [\Sigma X, \Sigma A^{\wedge |c|}]$$

for $c \in Q$ are defined as follows. Here Q is a set of basic commutators with a fixed total ordering.

(2.11) **Definition.** Let $F(z_1, z_2)$ be the set of brackets in the letters z_1, z_2 and let $L(z_1, z_2)$ be the free Lie algebra (non graded over $\mathbb{Q}$, generated by z_1, z_2. There is an obvious mapping $F(z_1, z_2) \to L(z_1, z_2)$ which we suppress from our notation. By the Poincaré-Birkhoff-Witt-theorem a subset $Q \subset F(z_1, z_2)$ can be chosen such that $\{z_1, z_2\} \cup Q$ is a basis of the $\mathbb{Q}$-vector space $L(z_1, z_2)$. In this case Q is said to be a set of basic commutators.

We choose an ordering of Q compatible with the weight, that is for $c \leq c'$ we have $|c| \leq |c'|$.

The functions h_c are now determined by the formula:

$$(2.12) \quad (i_1 + i_2)\alpha = i_1\alpha + i_2\alpha + \sum_{c \in Q} [i_1, i_2]_c \circ h_c(\alpha)$$

in $[\Sigma X, \Sigma A \vee \Sigma A]$. Summation is taken over the fixed total ordering of Q. The Hilton-Milnor-Theorem (see 4.7 [23]) shows that such functions h_c exist and are well defined by this formula.

Clearly, α is a Co-H-map that is $(i_1 + i_2)\alpha = i_1\alpha + i_2\alpha$, if and only if all Hilton-Hopf invariants $h_c(\alpha) = 0$ vanish. Thus (2.7) shows a connection with the James-Hopf invariants, in fact we deduce from (2.8) inductively.

(2.13) Corollary. *Let X be finite dimensional and let A be a Co-H-space. Then for* $\alpha \in [\Sigma X, \Sigma A]$ *the iterated James-Hopf invariants* $\gamma_{n_1}\gamma_{n_2}\cdots\gamma_{n_r}(\alpha)$ *determine all Hilton Hopf invariants* $h_c(\alpha)$.

Proof. $c_n(i_1, i_2)$ can be written as a sum of 'basic commutators' and we can apply (2.8) inductively. In this way we obtain from (2.8) a sum as in (2.12). By uniqueness of $h_c(\alpha)$ we have the conclusion of (2.13). //

Suspended versions of (2.13) were proved by M. Barratt [6] and J. M. Boardman-B. Steer [13].

(2.14) Proposition. *Let X, A, B be connected spaces and let* $\alpha \in [\Sigma X, \Sigma A]$ *and* $\beta \in [\Sigma X, \Sigma B]$. *If* $A = \Sigma A'$ *is a suspension we have*

$$\gamma_n(\alpha \cup \beta) = T \circ ((\gamma_n \alpha) \cup \beta^n)$$

where $\beta^n = \beta \cup \ldots \cup \beta$ *is the* n-*fold cup product and where* $T : \Sigma A^{\wedge n} \wedge B^{\wedge n} \to \Sigma(A \wedge B)^{\wedge n}$ *is the shuffle map.*

Proof. This follows from the homotopy commutative diagram

$$\begin{array}{ccc} X \wedge X & \xrightarrow{\ \bar{\alpha} \wedge 1\ } & J(A) \wedge X \\ & \searrow^{\overline{\alpha \wedge 1}} & \downarrow \rho \\ & & J(A \wedge X) \end{array}$$

where $\rho(a_1 \ldots a_r \wedge x) = (a_1 \wedge x) \ldots (a_r \wedge x)$. //

(2.15) Proposition. *Let* γ_n *be defined with respect to the lexicographical ordering from the left and let* $\alpha, \beta \in [\Sigma X, \Sigma A]$. *Then we have for* $n \geq 1$

$$\gamma_n(\alpha + \beta) = \gamma_n \alpha + \sum_{i=1}^{n-1} \gamma_{n-i}(\alpha) \cup \gamma_i(\beta) + \gamma_n(\beta)$$

Proof. Let μ be the multiplication on J(A). The proposition follows from the equation, $f_1, f_2 \in [\Sigma(JA \times JA), \Sigma A^{\wedge n}]$,

$$f_1 = \bar{g}_n(\Sigma\mu) = \sum_{i=0}^{n} \bar{g}_i q_1 \cup \bar{g}_{n-i} q_2 = f_2$$

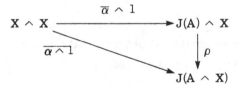

where $q_1, q_2 : \Sigma(JA \times JA) \to \Sigma JA$ are the projections. Now let

$$\pi_N : A^N = A \times \ldots \times A \to J(A)$$

be given by $\pi_N(x_1, \ldots, x_N) = x_1 \cdot \ldots \cdot x_N$ and let $\pi = \Sigma(\pi_N \times \pi_M)$. It is enough to prove $f_1 \circ \pi = f_2 \circ \pi$ for all N, M. In $[\Sigma(A^N \times A^M), \Sigma A^{\wedge n}]$ we have

$$f_2 \circ \pi = \sum_{i=0}^{n} \sum_{\substack{\#a=n-i \\ a\subset\overline{N}}}^{<} \sum_{\substack{\#b=i \\ b\subset\overline{M}}}^{<} (p_a q_1) \cup (p_b q_2)$$

On the other hand

$$f_1 \circ \pi = \sum_{\substack{B\subset\overline{N+M} \\ \#B=n}}^{<} p_B$$

We sum in the lexicographical ordering. For $a = \{a_1 < \ldots < a_r\} \subset \overline{N} = \{1, \ldots, N\}$ let

$$p_a = \Sigma A^N \to \Sigma A^{\wedge r}$$

be the projection $p_a(t, x_1, \ldots, x_N) = (t \wedge x_{a_1} \wedge \ldots \wedge x_{a_r})$.

It is a property of the lexicographical ordering that the sums $f_1 \circ \pi$ and $f_2 \circ \pi$ coincide. Thus (2.15) is proved. //

For $n, m \geq 1$ let $T(n, m)$ be the set of all partitions

$$A = (a^1, \ldots, a^m) \in \mathrm{Par}(n \cdot m)$$

of $\overline{nm}$ with $\#a^1 = \ldots = \#a^m = n$ and with $\mathrm{Min}(a^1) < \ldots < \mathrm{Min}(a^m)$. See (I. 2. 3). The function $0 : \mathrm{Par}(nm) \to S_{nm}$ is the shuffle permutation (see § 3).

(2.16) Theorem. Let γ_n be defined with respect to the lexicographical ordering from the left. Let X be finite dimensional and let Y be a Co-H-space. For $\alpha \in [\Sigma X, \Sigma Y]$ we have

$$\gamma_m \gamma_n(\alpha) = T_{n,m} \circ \gamma_{nm}(\alpha)$$

where

$$T_{n,m} = \sum_{A \in T(n,m)} T_{o(A)}$$

with T_σ for $\sigma \in S_{nm}$ as in (2.8).

Compare III (5.2).

§3. The Whitehead product of composition elements

Consider the compositions

$$(3.1) \quad \begin{cases} \xi\alpha : \Sigma A \xrightarrow{\alpha} \Sigma X \xrightarrow{\xi} Z \\ \eta\beta : \Sigma B \xrightarrow{\beta} \Sigma Y \xrightarrow{\eta} Z \end{cases}$$

where A, B, X, Y, Z are connected spaces. In this section we prove an expansion formula for the Whitehead product

$$[\xi\alpha, \eta\beta] \in [\Sigma A \wedge B, Z]$$

This formula yields a proof of the commutator formula I (2.6).

(3.2) **Proposition.** If α and β are Co-H-maps then

$$[\xi\alpha, \eta\beta] = [\xi, \eta] \circ (\alpha \# \beta) .$$

Proof. The proposition follows from

$$\begin{aligned} [\xi\alpha, \eta] &= [\xi \circ R(\Sigma\bar{\alpha}), \eta] \\ &= [\xi R, \eta] \circ (\Sigma\bar{\alpha} \wedge Y) \\ &= [\xi R, \eta] \circ ((\Sigma i) \wedge Y) \circ (\alpha \wedge Y) \\ &= [\xi R(\Sigma i), \eta] \circ (\alpha \wedge Y) = [\xi, \eta](\alpha \wedge Y) \end{aligned}$$

where $\Sigma\bar{\alpha} = (\Sigma i) \circ \alpha$ by use of Ganea's diagram in the proof of (2.7), here also (1) in the proof of (2.8). //

We have canonical shuffle permutations

$$0, 1 : \mathrm{Par}(n) \to S_n$$

to the permutation group S_n of $\bar{n}$. For a partition $a = (a^1, \ldots, a^r) \in \mathrm{Par}(n)$

of $\overset{\text{II}}{\bar{n}}$ we write as usual $a^i = \{a_1^i < \ldots < a_{e_i}^i\}$ where e_i is the number of elements of a^i. Then

$$0(a)(e_1 + \ldots + e_{i-1} + j) = a_j^i$$
$$1(a)(e_1 + \ldots + e_{i-1} + j) = a_{e_i - j+1}^i$$

for $j = 1, \ldots, e_i$ and $i = 1, \ldots, r$.

For ξ and η in (3.1) we have as in I (2.5) the iterated brackets

$$[\xi^n, \eta^m] = [[[\xi^n, \eta]], \eta^{m-1}] \in [\Sigma X^{\wedge n} \wedge Y^{\wedge m}, Z]$$

For $i = (i_1, \ldots, i_r)$ and $j = (j_1, \ldots, j_r)$ with $i_1 + \ldots + i_r = n$, $j_1 + \ldots + j_r = m$ let

$$T_{i,j}^{X,Y} : \Sigma X^{\wedge i_1} \wedge Y^{\wedge j_1} \wedge \ldots \wedge X^{\wedge i_r} \wedge Y^{\wedge j_r} \approx \Sigma X^{\wedge n} \wedge Y^{\wedge m}$$

be the obvious permutation of factors collecting X and Y coordinates respectively.

For a permutation $\sigma \in S_n$ let

$$T_\sigma^Y : Y^{\wedge n} \to Y^{\wedge n}$$

be the associated permutation of coordinates,

$$T_\sigma^Y(y_1, \ldots, y_n) = (y_{\sigma 1}, \ldots, y_{\sigma n}).$$

With this notation we can state the expansion formula

(3.3) Theorem. <u>Let X and Y be Co-H-spaces and let A and B be finite dimensional. Then we have in</u> $[\Sigma A \wedge B, Z]$ <u>the formula</u>

$$[\xi\alpha, \eta\beta] = \sum_{N \geq 1} \sum_{M \geq 1} R_{M,N}(\xi, \eta) \circ (\gamma_M(\alpha) \underline{\#} \gamma_N(\beta))$$

<u>where</u>

$$R_{M,N}(\xi, \eta) \in [\Sigma X^{\wedge M} \wedge Y^{\wedge N}, Z]$$

<u>is defined by</u>

$$R_{M,N}(\xi,\eta) = \sum_{\substack{a\in R(N),\ x\in Par(M)\\ |a| = |x| = r\geq 1}} (-1)^{M-r}\prod [[[\xi^{\# x^1}, \eta^{\# a^1}], \dots, [\xi^{\# x^r}, \eta^{\# a^r}]] \circ T_{x,a}$$

with $T_{x,a} = T^{X,Y}_{\# x, \# a} \circ (\Sigma T^X_{1(x)} \wedge T^Y_{0(a)})$.

Clearly if X and Y are spheres, $T_{x,a}$ is just a sign. The theorem is a generalization of the following Barcus-Barratt formula:

(3. 4) **Proposition.** Assume A is a Co-H-space and assume B has finite dimension. Then the Whitehead product of

$$\Sigma A \xrightarrow{\zeta} Z$$

$$\Sigma B \xrightarrow{\beta} \Sigma Y \xrightarrow{\eta} Z$$

satisfies the formula

$$[\zeta, \eta\beta] = \sum_{n\geq 1} [[\zeta, \eta^n] \circ (A \wedge \gamma_n(\beta)).$$

This formula was proved by Barcus-Barratt [5] in case A and B are spheres and by Rutter if A and B are suspensions. In fact only A needs to be a Co-H-space. In [7] we show that (3. 4) is a special case of a whole series of similar formulas for Whitehead products and Hopf invariants.

We derive from (3. 4) by (1. 7) and linearity of the Whitehead product

(3. 4)' **Corollary.** Let A and B be Co-H-spaces of finite dimension, then for (3. 1) we have

$$[\zeta\alpha, \eta\beta] = \sum_{\substack{N\geq 1\\ M\geq 1}} ((-1)^{M-1}[\zeta^M, \eta^N]) \circ (\gamma'_M(\alpha) \,\underline{\#}\, \gamma_N(\beta))$$

where $\gamma'_M(\alpha) = (\Sigma T^X_\tau) \circ \gamma_M(\alpha)$ with $\tau = \binom{1 \dots M}{M \dots 1} \in S_M$.

Unfortunately formula (3. 4) is not available for our intended expansion of $[\zeta\alpha, \eta\beta]$ since in §5 we are not allowed to assume that A or B is a Co-H-space. We therefore prove the more sophisticated version (3. 5) below from which we will deduce a proof of (3. 3).

(3.5) **Proposition.** Assume Y is a Co-H-space. If B has finite dimension the Whitehead product of

$$\Sigma A \xrightarrow{\zeta} Z$$

$$\Sigma B \xrightarrow{\beta} \Sigma Y \xrightarrow{\eta} Z$$

satisfies the formula

$$[\zeta, \eta\beta] = \sum_{n\geq 1} \sum_{\substack{a\in R(n)\\ r=|a|}} K_a$$

with

$$K_a = [[[\zeta, \eta^{\#a_1}], \ldots, [\zeta, \eta^{\#a_r}]] \circ T^{A,Y}_{(1),\#a} \circ ((\Sigma\tilde{\Delta}_a) \underline{\#} (\hat{T}^Y_{\sigma(a)} \circ \gamma_n(\beta)))$$

where $\tilde{\Delta}_a : A \to A^{\wedge |a|}$ is the reduced diagonal and $(1) = (1, \ldots, 1)$, $\hat{T} = \Sigma T$.

Proof of (3.4), (3.5). Let $Z = \Sigma A \vee \Sigma Y$ and let ζ, η be the inclusions of ΣA and ΣY into Z respectively. Let

(1) $\qquad \bar{g}_Y : \Sigma J(Y) \to \Sigma Y$

be the adjoint of the homotopy equivalence $J(Y) \simeq \Omega\Sigma Y$ in (2.1). For the adjoint $\bar{\beta} : B \to J(Y)$ of β we have the formula

(2) $\qquad [\zeta, \eta\beta] = [\zeta, \eta\bar{g}_Y] \circ (\Sigma A \wedge \bar{\beta})$,

since $\beta = \bar{g}_Y \circ (\Sigma\bar{\beta})$.

Now let $s : Y^n \to J_n(Y) \subset J(Y)$ be the quotient map with $s(y_1, \ldots, y_n) = y_1 \cdot \ldots \cdot y_n$. We used already several times the fact that the composition

$$\bar{g}_Y(\Sigma s) : \Sigma Y^n \to \Sigma J(Y) \to \Sigma Y$$

satisfies

(3) $\qquad \bar{g}_Y(\Sigma s) = p_1 + \ldots + p_n$

where $p_i : \Sigma Y^n \to \Sigma Y$ is the projection onto the i-th coordinate.

From the definition of the Whitehead product in (1.6) we derive that the mapping

$$\Sigma(A \times Y^n) \xrightarrow{\hat{s}} \Sigma(A \wedge J(Y)) \xrightarrow{[\zeta,\ \eta\bar{g}_Y]} \Sigma A \vee Y = Z$$

with $\hat{s} = (\Sigma A \wedge s)(\Sigma\pi)$ is the commutator

(4) $$U = \hat{s}^*[\sigma,\ \eta\bar{g}_Y] = -p_A - (p_1+\ldots+p_n) + p_A + (p_1+\ldots+p_n)$$

where now

$$p_A : \Sigma(A \times Y^n) \to \Sigma A \overset{\zeta}{\hookrightarrow} Z$$
$$p_i : \Sigma(A \times Y^n) \to \Sigma Y \overset{\eta}{\hookrightarrow} Z$$

are given by the corresponding projections onto the factors of the product $A \times Y^n$. On the other hand we have the homotopy equivalence

(5) $$\Sigma(A \wedge J(Y)) \xrightarrow{G} \bigvee_{r\geq 1} \Sigma A \wedge Y^{\wedge r},$$

which is given as in (2.6). Let j_r $(r \geq 1)$ be the inclusion of $\Sigma A \wedge Y^{\wedge r}$. Then G is the limit of

$$G_N = \sum_{r=1}^{N} j_r(A \wedge \bar{g}_r)$$

where $\bar{g}_r$ is defined as in (2.5), that is

$$\bar{g}_r = \bar{g}(\Sigma g_r) : \Sigma J(Y) \to \Sigma J(Y^{\wedge r}) \to \Sigma Y^{\wedge r}.$$

From the definition of g_r in (2.3) we see

(6) $$G\hat{s} = \sum_{r=1}^{n} \sum_{\substack{a \subset \bar{n} \\ \#a=r}}^{<} j_r \circ (\pi\bar{p}_a)$$

where $\pi\bar{p}_a : \Sigma(A \times Y^n) \to \Sigma A \wedge Y^{\wedge r}$ is given by

$$\pi\bar{p}_a(t,\ x,\ y) = (t,\ x,\ y_{a_1} \wedge \ldots \wedge y_{a_r})$$

for $a = \{a_1 < \ldots < a_r\}$, $x \in A$, $y \in Y^n$, $t \in I$.

The sum in (6) is taken in the fixed admissible order over all subsets a of $\bar{n} = \{1, \ldots, n\}$.

By use of the commutator rule (1.13) in $[\Sigma(A \times Y^n), Z]$ we now can 'collect' the summands of (4) (creating Whitehead products) in such a way that the sum obtained obviously factors over (6). This yields the result since $\mathfrak{s}^*$ is a monomorphism. First we get from lemma I (1.15)

$$U = -p_A - p_n - \ldots - p_1 + p_A + p_1 + \ldots + p_b = \tag{7}$$
$$= \sum_{i=n}^{1} \sum_{\substack{y \subset \bar{n} \\ \mathrm{Min}(y)=i}} (A, y)$$

with

$$(A, y) = [[p_A, p_{y_1}, \ldots, p_{y_{\#y}}].$$

Here $[\ ,\ ]$ denotes the commutator in $[\Sigma(A \times Y^n), Z]$. Thus the commutator of (A, y) and (A, x) is trivial if A is a co-H-space or if Y is a co-H-space and $y \cap x$ is non empty. Therefore proposition (3.4) is a consequence of (7) and (6). However, under the assumption of (3.5) the sum (7) is not yet in order $<$.

Collecting once more we obtain in lemma I (2.4) the formula

$$U = \sum_{y \subset \bar{n}}^{<} \sum_{a \in R(y)} (A, p)^a \tag{8}$$

where for $a = a^1 \ldots a^r$

$$(A, p)^a = [[(A, a^1), \ldots, (A, a^r)]$$

is the iterated commutator. A subset $y \subset \bar{n}$ yields the projection

$$\bar{p}_y : \Sigma(A \times Y^n) \to \Sigma(A \times Y^{\#y})$$

as in (6). From the definition in (0.17) we now see that

$$\sum_{a \in R(y)} (A, p)^a = \Big(\sum_{a \in R(\#y)} (A, p)^a \Big) \circ \bar{p}_y . \tag{9}$$

Therefore by use of (8) we get

$$(10) \qquad U = \overset{<}{\sum_{y \subset \bar{n}}} \Big(\sum_{a \in R(\# y)} (A, p)^{a} \Big) \circ \bar{p}_{y}$$

where we now sum in the fixed order over all subsets $y \subset \bar{n}$. As in (1.13) the commutator $(A, p)^{a}$ coincides with the composition

$$(A, p)^{a} = K_{a}\pi : \Sigma(A \times Y^{n}) \to \Sigma A \wedge Y^{\wedge n} \to Z.$$

Therefore the proposition in (3.5) follows from (6). //

Proof of (3.3). We use the same notation as in the proof of (3.5). Instead of (2) we now consider

$$(11) \qquad [\zeta\alpha, \eta\beta] = [\zeta\bar{g}_{X}, \eta\bar{g}_{Y}] \circ \Sigma(\overline{\alpha} \wedge \overline{\beta})$$

where ζ and η are the inclusions of ΣX and ΣY into $Z = \Sigma X \vee \Sigma Y$ respectively. As in (4) we obtain for the composition

$$\Sigma(X^{m} \times Y^{n}) \xrightarrow[\hat{s}]{} \Sigma(J(X) \wedge J(Y)) \xrightarrow[{[\zeta\bar{g}_{X}, \eta\bar{g}_{Y}]}]{} Z$$

the formula

$$(12) \qquad U = \hat{s}^{*}[\zeta\bar{g}_{X}, \eta\bar{g}_{Y}] = -q - p + q + p$$

where $p = p_{1} + \dots + p_{n}$, $q = q_{1} + \dots + q_{m}$ are sums of all projections

$$q_{i} : \Sigma(X^{m} \times Y^{n}) \to \Sigma X \hookrightarrow Z$$

$$p_{j} : \Sigma(X^{m} \times Y^{n}) \to \Sigma Y \hookrightarrow Z$$

$(i = 1, \dots, m)$, $(j = 1, \dots, n)$.

In the group $[\Sigma(X^{m} \times Y^{n}), Z]$ we derive as in (7) the formula

$$(13) \qquad [q, p_{j}] = \sum_{\substack{x \subset \bar{m} \\ \# x = r \geq 1}} (-1)^{r-1}[q_{x_{r}}, q_{x_{r-1}}, \dots, q_{x_{1}}, p_{j}]]$$

where we can sum in arbitrary order. Furthermore we have for $j = (j_{1}, \dots, j_{k})$ by use of (1.13)

$$(14) \qquad (q, j) = [[q, p_{j_{1}}, p_{j_{2}}, \dots, p_{j_{k}}] = \sum_{\substack{x \subset \bar{m} \\ \# x = r \geq 1}} (-1)^{r-1}[x, j]$$

with $\overset{\Pi}{[x,\, j]} = [[[q_{x_r}, \ldots, q_{x_1}, p_{j_1}]], p_{j_2}, \ldots, p_{j_k}].$

If we set $p_A = q$ in (7) we know from (8)

$$(15)\qquad U = \sum_{y\subset \bar{n}}^{<} \sum_{\substack{a\in R(y)\\ |a|=r}} [[(q, a^1), \ldots, (q, a^r)].$$

Using multilinearity, (1.13) and (14) we deduce

$$(16)\qquad U = \sum_{y\subset \bar{n}}^{<} \sum_{\substack{a\in R(y)\\ |a|=r}} \sum_{x^1,\ldots,x^r\subset \bar{m}} \varepsilon[[[x^1, a^1], \ldots, [x^r, a^r]]$$

with $\varepsilon = (-1)^{\#x^1+\ldots+\#x^r-r}$.

Here we sum only over all tuples $x^1, \ldots, x^r$ of non empty subsets of $\bar{m}$ which are disjoint. The sum over the indices $y \subset \bar{n}$ is taken in the fixed order. If we fix y, the remaining partial sum can be taken in arbitrary order. We therefore get

$$(17)\qquad U = \sum_{y\subset \bar{n}}^{<} \sum_{x\subset \bar{m}}^{<} \sum_{\substack{a\in R(y)\\ x^1\cup\ldots\cup x^r = x\\ r=|a|}} (-1)^{(\#x)-r}[[[x^1, a^1], \ldots, [x^r, a^r]]$$

where we take the sum over y and over x in the chosen order. As in (9) and (10) we get from (17) the result in (3.3) since we have by use of (17)

$$(18)\qquad U = \sum_{\substack{y\subset \bar{n}\\ \#y=N}}^{<} \sum_{\substack{x\subset \bar{m}\\ \#x=M}}^{<} R_{N,M}(\zeta, \eta) \circ (q_x \underline{\cup} p_y)$$

where

$$p_x : \Sigma(X^m \times Y^n) \to \Sigma X^{\wedge(\#x)}$$

$$p_y : \Sigma(X^m \times Y^n) \to \Sigma Y^{\wedge(\#y)}$$

are the projections onto the coordinates in $x \subset \bar{m}$ or $y \subset \bar{n}$ respectively. $R_{N,M}$ is defined as in (3.3). We use the fact that

$$\sum_{N\geq 1} \sum_{M\geq 1} g_M^{-X} \,\underline{\#}\, g_N^{-Y} : \Sigma(J(X) \wedge J(Y)) \to \bigvee_{M,N} \Sigma X^{\wedge M} \wedge Y^{\wedge N}$$

is an equivalence, compare (5) and (2. 6). //

§4. Proof of I (1. 13) and I (2. 6)

Let $L(x, y)$ be the non graded free Lie algebra over $\mathbb{Q}$ generated by x and y. We construct embeddings of non graded Lie algebras

(4. 1) $\psi : L(x, y) \subset L' = \mathrm{Hom}_{\mathbb{Q}}(T(z), L(z_1, z_2))$,

(4. 2) $\phi : L(x, y) \subset L'' = \mathrm{Hom}_{\mathbb{Q}}(T(u_1) \otimes T(u_2), L(u_1, u_2))$,

(4. 3) here $T(z) = H_*(J(S^n), \mathbb{Q})$

is the free graded tensor algebra generated by an element z of even degree n and $L(z_1, z_2)$ is the free graded Lie algebra over $\mathbb{Q}$ generated by z_1 and z_2 with degree $n = |z_1| = |z_2|$. $T(z)$ is a coalgebra with the diagonal

$$\Delta : T(z) \to T(z) \otimes T(z)$$

$$\Delta(z) = 1 \otimes z + z \otimes 1$$

which is an algebra homomorphism, thus

$$\Delta(z^n) = \sum_{k=0}^{n} \binom{n}{k} z^k \otimes z^{n-k} .$$

In (I. 3. 4) we saw that the $\mathbb{Q}$ vector space of degree zero homomorphisms $\mathrm{Hom}_{\mathbb{Q}}(C, \pi)$ is a non graded Lie algebra. Now we can define ψ to be the Lie algebra homomorphism with

$$\psi(x) = \hat{z} \otimes z_1, \quad \psi(y) = \hat{z} \otimes z_2$$

where $\hat{z} : T(z) \to \mathbb{Q}$ maps z to 1 and maps z^n to zero for $n \neq 1$, compare I (3. 6).

In a similar way we define the embedding ϕ: Let u_1 and u_2 be elements of degree n and m respectively where n and m is even. Then $T = T(u_1) \otimes T(u_2)$ is a graded coalgebra with the diagonal

$$\Delta : T \xrightarrow{\Delta \otimes \Delta} T(u_1) \otimes T(u_1) \otimes T(u_2) \otimes T(u_2) \overset{T}{\cong} T \otimes T$$

where τ is the switch of the terms in the middle. We have a coalgebra isomorphism

(4.4) $T(u_1) \otimes T(u_2) \cong H_*(J(S^n) \times J(S^m), \mathbb{Q})$.

The embedding ϕ is defined to be the Lie algebra homomorphism with

$$\phi(x) = \hat{u}_1 \otimes u_1, \quad \phi(y) = \hat{u}_2 \otimes u_2$$

where $\hat{u}_1, \hat{u}_2 : T \to \mathbb{Q}$ map $u_1 \otimes 1$ to 1 and $1 \otimes u_2$ to 1 respectively and map all other elements $u_1^n \otimes u_2^m$ to zero.

Looking at the images of basic commutators we see that ϕ and ψ are actually embeddings. We remark that

(4.5) $L(u_1, u_2) = \pi_*(\Omega(\Sigma S^n \vee \Sigma S^m)) \otimes \mathbb{Q}$.

For each N we have the mapping

(4.6) $\pi_N : P_N^n = (S^n)^N \to J(S^n)$

which is the restriction of the identification map π in (2.1). They induce mappings of Lie algebras (see (4.3) and (4.4))

(4.7)
$$L' \xrightarrow[s_N = \pi_{N*}^*]{} LP_N = \mathrm{Hom}(H_*(P_N^n, \mathbb{Q}), L(z_1, z_2))$$
$$L'' \xrightarrow[s_{NM} = (\pi_N \times \pi_M)_*^*]{} LP_{N,M} = \mathrm{Hom}(H_*(P_N^n \times P_M^m, \mathbb{Q}), L(z_1, z_2))$$

with the property

(4.8) Lemma. Let $f, g \in L'$. Then $f = g$ iff $s_N f = s_N g$ for all N. Similarly, let $f, g \in L''$. Then $f = g$ iff $s_{NM} f = s_{NM} g$ for all N, M.

For a Lie algebra L let $L \to \lambda L$ be a quotient map where λL is a nilpotent Lie algebra and assume that

(4.9)
$$\begin{cases} \lambda L(x, y) \overset{\tilde{\psi}}{\subset} \lambda L' \xrightarrow{s_N} LP_N \\ \lambda L(x, y) \underset{\tilde{\phi}}{\subset} \lambda L'' \xrightarrow[s_{NM}]{} LP_{NM} \end{cases}$$

are maps induced by (4.1), (4.2) and (4.7). For the proof of I (1.13) it is enough to prove that equation I (3.3) holds in the group $\exp(\lambda L(x, y))$. Similarly for the proof of I (2.6) it is enough to show equation I (3.3)' in the group $\exp(\lambda L(x, y))$. Because of (4.8) we have to prove that the images of these equations under (4.9) are in fact equations in LP_N, LP_{NM} for all N, M.

Now let P be a product of spheres of even dimension $n_i \geq 2$ with the product cell structure given by the cell decomposition $S^n = e^n \cup \{\text{basepoint}\}$ of S^n. For each cell e in P we have the canonical projection $p_e : P \to S^{|e|}$ which is of degree 1 on e. We denote the cohomology class given by p_e by e.

(4.10) **Proposition.** <u>There is an unique isomorphism of rational nilpotent groups</u>

$$\exp \operatorname{Hom}(H_*(P, Q), \pi_*(\Omega Z) \otimes \mathbb{Q}) \underset{\sigma}{\cong} [\Sigma P, Z]_{\mathbb{Q}}$$

<u>with</u> $\sigma(e \otimes \alpha) = \bar{\alpha} \circ (\Sigma p_e)$ <u>for all cells</u> e <u>in</u> P <u>and</u> $\alpha \in \pi_{|e|}(\Omega Z) \otimes \mathbb{Q}$, $\bar{\alpha} \in [\Sigma S^{|e|}, Z]_{\mathbb{Q}}$ <u>is the adjoint of</u> α.

See I (3.1) and I (3.6). The proposition follows since the group structure exp satisfies the <u>same</u> commutator rule as we found in (1.13).

If n is even we know that for $\alpha \in \pi_n(\Omega Z) \otimes \mathbb{Q}$ the bracket $[\alpha, \alpha]$ is trivial. Therefore for all $u, v \in H^n(P, \mathbb{Q})$ the bracket

$$[u \otimes \alpha, v \otimes \alpha] = (u \cup v) \otimes [\alpha, \alpha] = 0$$

is trivial. Thus the elements $\{u \otimes \alpha \mid u \in H^n(P, \mathbb{Q})\}$ generate an abelian subgroup of

$$\exp(H_*(P, \mathbb{Q}), \pi_*(\Omega Z) \otimes \mathbb{Q}) .$$

This proves

(4.11) **Lemma.** <u>Let</u> $u : H_n(P, \mathbb{Q}) \to \mathbb{Q}$ <u>be a homomorphism. Then the isomorphism</u> σ <u>in (4.10) satisfies</u>

$$\sigma(u \otimes \alpha) = \sum_e (u(e) \cdot \bar{\alpha}) \circ (\Sigma p_e)$$

where we sum over all cells e of dimension n in P. For summation we can choose any ordering.

Proof of I (1.13) and I (2.6). Let $p_i : P^n_N \to S^n$ be the projection onto the i-th coordinate $(i = 1, \ldots, N)$. We define

$$\alpha = (\Sigma p_1) + \ldots + (\Sigma p_N) : \Sigma P^n_N \to \Sigma S^n.$$

Let $i_1, i_2 : \Sigma S^n \hookrightarrow \Sigma S^n \vee \Sigma S^n$ be the inclusion. By use of (4.11) we get for the homomorphism in (4.9) and (4.10)

(1) $$\sigma s_N \tilde{\psi} x = \sigma s_N (\hat{x} \otimes z_1)$$
$$= \sigma (\pi^*_N \hat{x} \otimes z_1)$$
$$= i_1 \circ \alpha$$

(2) $$\sigma s_N \tilde{\psi} y = i_2 \circ \alpha$$

with $c_1(i_1, i_2) = i_1 + i_2$ we derive from (2.8) the equation

(3) $$\sigma s_N \tilde{\psi}(x\ y) = i_1 \alpha + i_2 \alpha$$
$$= \sum_{k \geq 1} c_k(i_1, i_2) \circ \gamma_k(\alpha)$$

On the other hand

(4) $$\sigma s_N \tilde{\psi}(c_k(x, y)/k!) =$$

(5) $$= \sigma s_N((\hat{x}^k/k!) \otimes c_k(z_1, z_2)),$$ compare I (3.8),

(6) $$= \sigma (\frac{\pi^*_N(\hat{x})^k}{k!} \otimes c_k(z_1, z_2))$$

(7) $$= \sum_e c_k(i_1, i_2) \circ (\Sigma p_e)$$

(8) $$= c_k(i_1, i_2) \circ \gamma_k(\alpha).$$

(4) and (5) follow from the definition of $\tilde{\psi}$ and s_N. (6) is a consequence of (4.11), we sum in (6) over all cells of P^n_N of dimension $n \cdot k$ in the chosen order. From the definition of γ_k and α we derive (8). From (3) and (8) and the remarks following (4.9) we deduce the proposition of

I (1.13).

In the same way we prove the proposition of I (2.6). Let

$$i_1\alpha,\ i_2\beta : \Sigma P_{N,M} \to \Sigma S^n \vee \Sigma S^m$$

be given by $\alpha = (\Sigma p_1 + \ldots + \Sigma p_N)$ and $\beta = (\Sigma q_1 + \ldots + \Sigma q_M)$ where p_i and q_j are the projections onto S^n and S^m respectively. Then we know as above

$$\begin{aligned}\sigma s_{NM}\tilde{\phi}(x^{-1}y^{-1}xy) &= -i_1\alpha - i_2\beta + i_1\alpha + i_2\beta \\ &= [i_1\alpha,\ i_2\beta] \circ (\Sigma\tilde{\Delta}),\end{aligned}$$

where we use the formula in (1.12). From (3.3) we now derive I (2.6). //

§5. Decomposition of suspensions and groups of homotopy classes

We study here the group $[\Sigma X, Z]$ in case ΣX is decomposable as a wedge

$$\Sigma X \simeq \bigvee_i \Sigma Y_i$$

where all Y_i are Co-H-spaces.

In the following all spaces X, Z, Y_i are path connected. Let $Y = (Y_i \mid i \in J)$ be a family of Co-H-spaces and let

(5.1) $\hat{Y} = \{\hat{Y}_i = Y_{i_1} \wedge \ldots \wedge Y_{i_k} \mid i \in \hat{J}\}$

be its associated family of smash products where $i = i_1 \ldots i_k \in \hat{J} = \mathrm{Mon}(J)$ runs through all words with letters $i_1, \ldots, i_k \in J$, $k \geq 0$. From $\hat{Y}$ we obtain families of groups

$$[\Sigma\hat{Y}, Z] = ([\Sigma\hat{Y}_i, Z] \mid i \in \hat{J}),$$

$$[\Sigma X, \Sigma\hat{Y}] = ([\Sigma X, \Sigma\hat{Y}_i] \mid i \in \hat{J}).$$

The Whitehead product $[\ ,\ ]$, the geometric cup product $\underline{\cup}$ and the Hopf invariants γ_r are additional structures on these families of groups. Again for the definition of γ_r we choose an admissible ordering of $P(\mathbb{N})$.

$$\begin{aligned}
&[\ ,\ ] : [\Sigma\hat{Y}, Z] \times [\Sigma\hat{Y}, Z] \to [\Sigma\hat{Y}, Z] \\
(5.2)\quad &\underline{\cup} : [\Sigma X, \Sigma\hat{Y}] \times [\Sigma X, \Sigma\hat{Y}] \to [\Sigma X, \Sigma\hat{Y}] \\
&\gamma_r : [\Sigma X, \Sigma\hat{Y}] \to [\Sigma X, \Sigma\hat{Y}]
\end{aligned}$$

Let $i = |x| = |\alpha|$ be the degree of $\alpha \in [\Sigma\hat{Y}_i, Z]$ or $x \in [\Sigma X, \Sigma\hat{Y}_i]$ with $i \in \hat{J}$ and let i' be the degree of α', x'. Then $[\alpha, \alpha']$ and $x \underline{\cup} x'$ have degree ii' and $\gamma_r(x)$ has degree $i^r = i \dots i$, the r-fold product of i.

If Z is an H-space the pairing [,] is trivial and if X is a Co-H-space $\underline{\cup}$ is trivial.

(5.3) **Assumption (*).** Let X and ΩZ be connected and let X be finite dimensional or Z be finite codimensional.

The assumption implies that iterated Whitehead products, iterated cup products or Hopf invariants γ_r are trivial for sufficiently large r.

We are interested in the subgroup of the group $[\Sigma X, Z]$ which is generated by all mappings

$$\Sigma X \to \Sigma\hat{Y}_i \to Z, \quad i \in \hat{J},$$

factoring over a space of the family $\Sigma\hat{Y}$.

This subgroup is the image of the natural group homomorphism

$$(5.4)\quad \rho : [\Sigma X, \Sigma\hat{Y}, Z] \to [\Sigma X, Z]$$

which we now describe. The disjoint union

$$E = \bigcup_{i \in \hat{J}} [\Sigma X, \Sigma\hat{Y}_i] \times [\Sigma\hat{Y}_i, Z]$$

is the set of generators for the group

$$[\Sigma X, \Sigma\hat{Y}, Z] = FG(E)/\sim$$

The relation $\sim$ in the free group FG(E) on E is generated by the relations (i) ... (iv) below. On generators $(x, z) \in E$ the homomorphism ρ in (5.4) is given by composition

(5.5) $\rho(x, \alpha) = \alpha \circ x.$

The relations on FG(E) are defined in terms of generators $\in E$ as follows:

(i) $(x, \alpha) \cdot (y, \alpha) \sim (x + y, \alpha)$

(ii) $(x, \alpha) \cdot (x, \beta) \sim (x, \alpha + \beta) \cdot \prod_{n \geq 2} (\gamma_n x, c_n(\alpha, \beta))$

(iii) $(x, \alpha)^{-1}(y, \beta)^{-1}(x, \alpha)(y, \beta) \sim$

$$\prod_{n \geq 1} \prod_{m \geq 1} (\gamma_m x \underline{\cup} \gamma_n y, R_{m,n}(\alpha, \beta))$$

(iv) $(\eta \circ x, \alpha) \sim (x, \alpha \circ \eta)$ for $\eta \in [\Sigma\hat{Y}_{|x|}, \Sigma\hat{Y}_{|\alpha|}]$

for $x, y \in [\Sigma X, \Sigma\hat{Y}]$ and $\alpha, \beta \in [\Sigma\hat{Y}, Z]$. In (i) we have $|x| = |y| = |\alpha|$ and in (ii) $|x| = |\alpha| = |\beta|$.

These relations look similar to those of the exponential group I (3.7), however their meaning is slightly more general. The Zassenhaus term $c_n(\alpha, \beta)$ here is that of (2.8) and the term $R_{n,m}(\alpha, \beta)$ is defined in (3.3). In fact it is an easy consequence of (2.8) and (3.3) that ρ is a well-defined homomorphism of groups. Here we make use of the commutator equation

(5.6) $-\alpha \circ x - \beta \circ y + \alpha \circ x + \beta \circ y = [\alpha \circ x, \beta \circ y] \circ (\Sigma\tilde{\Delta}),$

compare the proof of (1.12).

We point out that the group $[\Sigma X, \Sigma\hat{Y}, Z]$ depends only on the structure maps (5.2) and on composition with elements in $[\Sigma\hat{Y}_i, \Sigma\hat{Y}_j]$ in (iv). Clearly $[\Sigma X, \Sigma\hat{Y}, Z]$ is an abelian group as is $[\Sigma X, Z]$ if X is a Co-H-space or Z is an H-space.

We now describe a condition under which the homomorphism ρ in (5.4) is actually an isomorphism.

(5.7) **Definition.** Let $Y = (Y_i | i \in J)$ be a family of Co-H-spaces. We say a suspension ΣX has a $\hat{Y}$-decomposition, if there exist $i_n \in \hat{J}$, $(1 \leq n \leq N \leq \infty)$, and mappings $\xi_n : \Sigma X \to \Sigma\hat{Y}_{i_n}$ so that

$$\xi = \sum_{n=1}^{N} j_n \circ \xi_n : \Sigma X \to \bigvee_{n=1} \Sigma\hat{Y}_{i_n}$$

is a homotopy equivalence. Here j_n denotes the inclusion of $\Sigma\hat{Y}_{i_n}$ into

II
the wedge.

(5.8) **Remark.** If $N = \infty$ we assume that for $n \to \infty$ the connectivity of $\hat{Y}_{i_n}$ also tends to infinity. ξ is the limit of the finite sums in this case.

(5.9) **Theorem.** Let X be finite dimensional or Z be finite codimensional. If ΣX admits a $\hat{Y}$-decomposition,

$$\rho : [\Sigma X, \Sigma\hat{Y}, Z] \cong [\Sigma X, Z]$$

is an isomorphism of groups.

Clearly if a $\hat{Y}$-decomposition ξ of ΣX is given we have a bijection of sets

(5.10) $$\xi^* : [\Sigma X, Z] \approx \mathop{\times}_{n=1}^{N} [\Sigma \hat{Y}_{i_n}, Z]$$

Because of (5.8) this is a finite product.

(5.11) **Example.** The suspension of a finite product of Co-H-spaces Y_i has the $\hat{Y}$-decomposition

$$\Sigma(\mathop{\times}_{i=1}^{N} Y_i) = \bigvee_{\substack{1 \le i_1 < \ldots < i_r \le N \\ r \ge 1}} \Sigma Y_{i_1} \wedge \ldots \wedge Y_{i_r},$$

a result of D. Puppe, Math. Z. 69, 299-344 (1958). The loop space $\Omega\Sigma Y$ for a Co-H-space Y has the decomposition

$$\Sigma(\Omega\Sigma Y) \simeq \bigvee_{n \ge 1} \Sigma Y^{\wedge n}.$$

Proof of (5.9). First we observe that all cosets of $G = [\Sigma X, \Sigma\hat{Y}, Z]$ are represented by words

(1) $(x_1, \alpha_1) \cdot \ldots \cdot (x_r, \alpha_r)$ of generators $(x_i, \alpha_i) \in E$. This we know since by (i) we have

(2) $(x, \alpha)^{-1} \sim (-x, \alpha)$ and $(x, 0) \sim 1$.

We say a coset $g \in G$ has length $L(g) \le r$ if it contains a word of generators as in (1) of length r. We now prove inductively the

(3) Proposition. Each $g \in G$ contains an element of the form

$$\prod_{n=1}^{N} (\xi_n, \alpha_n)$$

with $\alpha_n \in [\Sigma\hat{Y}_{i_n}, Z]$ and with ξ_n as in (5.7).

First we see this for $L(g) = 1$. In this case we know for a generator $(x, \alpha) \in g$ from (5.10) that

$$(4) \qquad x = \sum_{n=1}^{N} \beta_n \xi_n, \ \beta_n \in [\Sigma\hat{Y}_{i_n}, \Sigma\hat{Y}_{|x|}].$$

Thus we get from (i) and (iv) the relations:

$$(5) \qquad (x, \alpha) \sim (\sum_{n=1}^{N} \beta_n \xi_n, \alpha)$$

$$\sim \prod_n (\beta_n \xi_n, \alpha)$$

$$\sim \prod_n (\xi_n, \alpha\beta_n)$$

Assume now (3) is proven for all $g' \in G$ with $L(g') \leq L$, $L \geq 1$. Then clearly $g \in G$ with $L(g) = L + 1$ is representable by $\Gamma \in g$ with

$$(6) \qquad \Gamma = (\prod_{i=n}^{N} (\xi_i, \alpha_i)) \cdot (\prod_{i=n}^{N} (\xi_i, \alpha_i'))$$

By inductive use of the relations (i), ..., (iv) we now prove

$$\Gamma \sim \prod_{i=n}^{N} (\xi_i, \alpha_i'')$$

for certain elements α_i''.

We may assume that all spaces Y_i are CW-complexes with skeleta Y_i^r and $Y_i^0 = *$. We say $(x, \alpha) \in E$ has connectivity $0(x, \alpha) \geq r$ if α factors up to homotopy over the quotient map π

$$\alpha : \Sigma\hat{Y}_{|\alpha|} \xrightarrow{\pi} \Sigma(\hat{Y}_{|\alpha|}/\hat{Y}^{r-1}_{|\alpha|}) \rightarrow Z$$

Clearly $0(x, \alpha) \geq 1$ for all $(x, \alpha) \in E$ since we assume all Y_i to be connected. Since $\Sigma\hat{Y}_i = \Sigma Y_{i_1} \wedge \ldots \wedge Y_{i_r}$ is r-connected we know that each (x, α) with $|x| = |\alpha| = i$ has connectivity $0(x, \alpha) \geq r$.

For the terms in the relations (ii) and (iii) we obtain by use of

Π

properties of the Whitehead product, see (3. 4):

(7) For $0(x, \alpha) \geq r$ and $0(x, \beta) \geq s$ we have
$0(\gamma_n(x), C_n(\alpha, \beta)) \geq r + s$ for all $n \geq 2$.

(8) For $0(x, \alpha) \geq r$ and $0(y, \beta) \geq s$ we have
$0(\gamma_m(x) \underline{\cup} \gamma_n(y), R_{m,n}(\alpha, \beta)) \geq r + s$ for all $m, n \geq 1$.

Furthermore, we obtain by the cellular approximation theorem

(9) If $0(x, \alpha) \geq r$ then for all terms in (5) we have
$0(\xi_n, \alpha\beta_n) \geq r$, $n = 1, \ldots, N$.

If for Γ in (6) we know that $0(\xi_i, \alpha_i) \geq r$ and $0(\xi_i, \alpha_i') \geq s$ for all $i = 1, \ldots, N$ then we can derive from (ii) and (iii) a relation

$$(10) \qquad \Gamma \sim \left(\prod_{i=1}^{N} (\xi_i, \alpha_i + \alpha_i')\right) \cdot \prod_{\lambda \in \Lambda} (a_\lambda, b_\lambda)$$

where $0(a_\lambda, b_\lambda) \geq r + s$ and where Λ is a certain ordered index set $(a_\lambda, b_\lambda) \in E$. By use of (5) and (9) we now see that

$$(11) \qquad (a_\lambda, b_\lambda) \sim \prod_{n=1}^{N} (\xi_n, b_\lambda, \beta_n^\lambda)$$

with $0(\xi_n, b_\lambda \beta_n^\lambda) \geq r + s$.

Now using the collecting process in (10) repeatedly we arrive by induction at an equivalence

$$(12) \qquad \Gamma \sim \prod_{i=1}^{N} (\xi_i, \alpha_i'')$$

since the connectivity of the terms (a_λ, b_λ) becomes bigger at each step. By (5. 3) the process is finite.

The proposition (5. 9) is now easily verified. Clearly ρ in (5. 4) is surjective. ρ is also injective since for $g \in \operatorname{Ker} \rho$ we have by (3) an element

$$\prod_{n=1} (\xi_n, \alpha_n) \in g$$

and thus

$$\rho(g) = \sum_{n=1}^{N} \alpha_n \circ \xi_n = 0$$

Since ξ **is a homotopy equivalence we see that** $\alpha_n = 0$ **and thus** $g = 1$ **by** (2). //

III. HOMOTOPY OPERATIONS ON SPHERES

We consider the generalized homotopy groups $[\Sigma X, \Sigma S^n]$ of spheres ΣS^n where ΣX is the suspension of a finite dimensional polyhedron X. These groups have additional structure, namely Whitehead products [,]; cup products $\cup$, James-Hopf-invariants γ_n and composition $\circ$. We exhibit the distributivity properties of these homotopy operations. In other words we give explicit formulas for

$$[\alpha \circ \alpha', \beta \circ \beta']$$

$$(\alpha + \alpha') \circ \beta$$

$$\alpha \cup \beta - (-1)^{|\alpha||\beta|} \beta \cup \alpha$$

$$\gamma_n(\alpha) \cup \gamma_m(\alpha)$$

$$\gamma_n \gamma_m(\alpha),\ \gamma_n(\alpha + \beta),\ \gamma_n(\alpha \cup \beta),$$

$$\gamma_n(\alpha \circ \beta)$$

In this chapter the Hopf invariants γ_n are always defined with respect to the lexicographical ordering from the left.

§1. Spherical Whitehead products and commutators

We call a mapping f <u>spherical</u> if f maps into a sphere. The following spherical Whitehead products are of particular importance.

(1.1) **Facts.** The generator $j_n \in \pi_{n+1}(\Sigma S^n) \cong \mathbb{Z}$ has the following Whitehead products.

$[j_n, j_n]$ has infinite order if n is odd, is trivial if $n = 0, 2, 6$ and has order 2 otherwise.

$[[j_n, j_n], j_n]$ has order 3 if n is odd > 1 and is trivial otherwise.

All iterated brackets $[j_n, \ldots, j_n]_c$ with more than 3 factors are trivial.

See [23] and the paper of Adams: Ann. Math. 72 (1960) 20-104, and see the paper of Liulevicius: Proc. N. A. S. 46 (1960) 978-81.

In the group $[\Sigma X, Y]$ the commutator

$$(\alpha, \beta) = -\alpha - \beta + \alpha + \beta$$

can be expressed by a Whitehead product

(1.2) $(\alpha, \beta) = [\alpha, \beta] \circ (\Sigma\tilde{\Delta})$.

(1.3) **Proposition.** Let X be a finite dimensional space and $\alpha, \beta, \gamma \in [\Sigma X, \Sigma S^n]$. The commutators satisfy the formulas $(j = j_n)$.

$$\begin{aligned}(\alpha, \beta) &= [j, j](\alpha \cup \beta) + \\ &\quad [[j, j], j](\alpha \cup \gamma_2\beta - (\gamma_2\alpha) \cup \beta), \\ ((\alpha, \beta), \gamma) &= [[j, j], j](\alpha \cup \beta \cup \gamma)\end{aligned}$$

and all iterated commutators of length ≥ 4 vanish.

By (1.2) proposition (1.3) is a consequence of the following formulas for Whitehead products which are special cases of the distributivity laws exhibited in chapter II.

(1.4) **Proposition.** Let A and B be finite dimensional. For $\alpha \in [\Sigma A, \Sigma S^n]$ and $\beta \in [\Sigma B, \Sigma S^n]$ we have the formula in $[\Sigma A \wedge B, \Sigma S^n]$ (with $j = j_n$):

$$\begin{aligned}[\alpha, \beta] = &[j, j] \circ (\alpha \,\underline{\#}\, \beta) \\ &+ ((-1)^{n-1}[j, [j, j]]) \circ ((\gamma_2\alpha) \,\underline{\#}\, \beta) \\ &+ [[j, j], j] \circ (\alpha \,\underline{\#}\, \gamma_2\beta).\end{aligned}$$

We are allowed to replace $\underline{\#}$ by $\#$ in this formula, see (3.1).

Proof. Because of (1.1) this follows as a special case of II (3.3), where we consider $[\alpha, \beta] = [j\alpha, j\beta]$. //

(1.5) **Corollary.** Let A_i be finite dimensional, $i = 1, 2, \ldots$. For $\alpha_i \in [\Sigma A_i, \Sigma S^n]$ there is the formula $(j = j_n)$:

III

$$[[\alpha_1, \alpha_2], \alpha_3] = [[j, j], j] \circ (\alpha_1 \underline{\#} \alpha_2 \underline{\#} \alpha_3)$$

and all iterated Whitehead products of more than three factors α_i vanish.

Proof. For $[\alpha_1, \alpha_2]$ we use (1. 4). The linearity of the Whitehead product II (3. 3) and (1. 1) above together yield the result. //

(1. 6) Proposition. Let X and Y be Co-H-spaces and let A and B be finite dimensional. For

$$\Sigma A \xrightarrow{\alpha} \Sigma X \xrightarrow{\xi} \Sigma S^n$$

$$\Sigma B \xrightarrow{\beta} \Sigma Y \xrightarrow{\eta} \Sigma S^n$$

we have the formula in $[\Sigma A \wedge B, \Sigma S^n]$

$$[\xi\alpha, \eta\beta] = [\xi, \eta] \circ (\alpha \underline{\#} \beta) + (-[\xi, [\xi, \eta]]) \circ ((T_X \circ \gamma_2 \alpha) \underline{\#} \beta) + [[\xi, \eta], \eta] \circ (\alpha \underline{\#} \gamma_2 \beta)$$

where $T_X : \Sigma X \wedge X \to \Sigma X \wedge X$ interchanges the two factors X, $T_X(t, x, y) = (t, y, x)$.

If $\xi = \eta = j_n$, this is exactly the proposition of (1. 4).

Proof. By (1. 5) this is a special case of II (3. 3). //

Moreover we obtain now as a special case of II (2. 8)

(1. 7) Proposition. Let A be finite dimensional and let X be a Co-H-space. For $\alpha \in [\Sigma A, \Sigma X]$ and $x, y \in [\Sigma X, \Sigma S^n]$ we have

$$(x+y) \circ \alpha = x\alpha + y\alpha + [x, y] \circ \gamma_2(\alpha) + c_3(x, y) \circ \gamma_3(\alpha)$$

where

$$c_3(x, y) = [[x, y], x]T^X_{213} + [[x, y], y](T^X_{312} + T^X_{213}).$$

Here $T = T^X_{ijk} : \Sigma X \wedge X \wedge X \to \Sigma X \wedge X \wedge X$

is the shuffle, mapping $t \wedge x_1 \wedge x_2 \wedge x_3$ to $t \wedge x_i \wedge x_j \wedge x_k$ for $t \in S^1$. Since T is just a sign if X is a sphere we obtain

(1.8) **Corollary.** Let A be finite dimensional and let $\alpha \in [\Sigma A, \Sigma S^n]$, $n \geq 1$. For $x, y \in \pi_{n+1}(\Sigma S^m)$ we have

$$(x+y) \circ \alpha = x\alpha + y\alpha + [x, y] \circ \gamma_2(\alpha) + c_3(x, y) \circ \gamma_3(\alpha)$$

where

$$c_3(x, y) = (-1)^n[[x, y], x] + ((-1)^n + 1)[[x, y], y]$$

If A is a Co-H-space, the term involving $\gamma_3(\alpha)$ is trivial.

(1.9) **Corollary.** Let A be finite dimensional, $\alpha \in [\Sigma A, \Sigma S^n]$ and let $j = j_n$ be the identity of ΣS^n. Then for $k \in \mathbb{Z}$

$$(kj)\alpha = k \cdot \alpha + (\frac{k(k-1)}{2}[j, j])\gamma_2(\alpha)$$
$$- (\frac{k(k-1)}{2}[[j, j], j])\gamma_3(\alpha)$$

If A is a Co-H-space, the term involving $\gamma_3(\alpha)$ is trivial.

Proof of (1.8), (1.9). If A is a Co-H-space the term involving $\gamma_3(\alpha)$ vanishes as we can see by (4.5). Moreover (1.9) follows by induction from (1.8), take $x = j$ and $y = kj$. //

For A a sphere, (1.9) was originally proved by P. J. Hilton in [23] in terms of the Hilton-Hopf invariant. That the correction term involving $\gamma_3(\alpha)$ in (1.8) vanishes when A is a sphere was first obtained by I. M. James and P. J. Hilton as remarked in the footnote on page 168 of [23].

§ 2. Spherical Hopf-invariants

We define functions

(2.1) $\lambda^r_n : [\Sigma X, \Sigma S^r] \to [\Sigma X, \Sigma S^{nr}]$

by

$$\lambda_n^r(\alpha) = \begin{cases} \gamma_n(\alpha) & \text{if } nr \text{ is even} \\ \gamma_{n-1}(\alpha) \cup \alpha & \text{if } nr \text{ is odd.} \end{cases}$$

where γ_n is the James-Hopf invariant with respect to the lexicographical ordering of $P(\mathbb{N})$ from the left. Thus for the definition of λ_n^r only James-Hopf invariants of <u>even</u> degree nr are involved.

We show that all James-Hopf invariants can be described in terms of the functions λ_n^r.

Let Λ_n^r be the set of functions

$$\lambda : [\Sigma X, \Sigma S^r] \to [\Sigma X, \Sigma S^{nr}]$$

given by

$$(2.2) \quad \lambda(\alpha) = \lambda_n^r(\alpha) + (s[j, j]) \circ \lambda_{2n}^r \alpha$$
$$(t[[j, j], j]) \circ \lambda_{3n}^r \alpha$$

where $s, t \in \mathbb{Z}$ and $j = j_{nr}$. Let Γ_n^r be the set of the James-Hopf invariants

$$(2.3) \quad \overset{<}{\gamma}_n : [\Sigma X, \Sigma S^r] \to [\Sigma X, \Sigma S^{nr}]$$

where $<$ varies over all admissible orderings of $P(\mathbb{N})$.

(2.4) Theorem. <u>If</u> X <u>is finite dimensional we have</u> $\Gamma_n^r \subset \Lambda_n^r$.

Clearly, if nr is even, Λ_n^r contains exactly two elements since $[j, j]$ is an element of order ≤ 2 by (1.1).

Remark. Composition with the suspension $[\Sigma X, \Sigma S^{nr}] \to [\Sigma^2 X, \Sigma^2 S^{nr}]$ yields a set $\Sigma\Lambda_n$ with exactly one element $\Sigma\lambda_n$. $\Sigma^{n-1}\lambda_n$ is the function considered by Boardman-Steer [13].

For several of the following proofs and for the proof of (2.4) we need a crucial lemma.

(2.5) Lemma. <u>Let</u> M <u>be a finite set and let</u> $G = FG(M)/\sim$ <u>be a group generated by</u> M <u>such that all iterated commutators of length</u> ≥ 4

are trivial. Let $<$ and $\dashv$ be two orderings on M, then in G we have the equation

$$\prod_{m\in M}^{<} m = \left(\prod_{m\in M}^{\dashv} m\right) \cdot \left(\prod_{P} [m, m']\right) \cdot \left(\prod_{Q} [[m, m'], m'']\right).$$

P is the set of all pairs (m, m') with ($m < m'$ and $m' \dashv m$) and Q is the set of all triples (m, m', m") with ($m < m'$ and $m' \dashv m$) such that ($m' < m''$ and $m'' \dashv m$) or ($m \dashv m''$). The products over P and Q can be taken in arbitrary order.

Because of (1. 3) this lemma is valid for any group $[\Sigma A, \Sigma S^n]$ where A is finite dimensional. Furthermore we will make use of the following properties of the reduced product. Let

(2. 6) $\pi = \pi_N : (S^n)^N \to J(S^n)$

be the quotient map with

$$\pi(x_1, \ldots, x_N) = x_1 \cdot \ldots \cdot x_N$$

and let $p_i : \Sigma (S^n)^N \to \Sigma S^n$ be the projection onto the i-th coordinate of the product $(S^n)^N = S^n \times \ldots \times S^n$. For any subset $a = \{a_1 < \ldots < a_r\} \subset \{1, \ldots, N\}$ we then have the projection

(2. 7) $p_a = p_{a_1} \cup \ldots \cup p_{a_r} : \Sigma(S^n)^N \to \Sigma S^{nr}$

The adjoint $\bar{g}_r : \Sigma J(S^n) \to \Sigma S^{nr}$ of the mapping $\overset{<}{g}_r$ in II (2. 3) has the property

(2. 8) Lemma. $\bar{g}_r(\Sigma\pi) = \overset{<}{\sum\limits_{a\subset\bar{N}}} p_a$ where we sum in the order $<$ over all subsets a of $\bar{N} = \{1, \ldots, N\}$ with r elements.

The sum is taken in the group $[\Sigma(S^n)^N, \Sigma S^n]$. The elements p_a and p_b have trivial commutator if a and b are not disjoint, the commutator is by (1. 3):

(2. 9) $(p_a, p_b) = [j, j](p_a \cup p_b)$

where $p_a \cup p_b = (\varepsilon_{a,b})^n p_{a\cup b}$. Here $\varepsilon_{a,b}$ is the shuffle sign of the partition (a, b) of $a \cup b$; compare II (1.13).

Moreover we use

(2.10) **Lemma.** Let $\alpha, \beta \in [\Sigma J(S^n), Z]$ and for all N let

$$(\Sigma\pi_N)^*\alpha = (\Sigma\pi_N)^*\beta .$$

Then $\alpha = \beta$.

Proof of (2.4). If nr is odd we show in (4.4) that $\gamma_n \in \Lambda_n^r$. Therefore $\gamma_n \in \Lambda_n^r$ for all n.

We now show that for two admissible orderings $<$, $\dashv$ on $P(\mathbb{N})$ we have $s, t \in \mathbb{Z}$ with

$$(1)\qquad -\overset{\dashv}{\gamma}_n(\alpha) \overset{<}{+} \gamma_n(\alpha) = (s[j, j]) \circ \gamma_{2n}(\alpha) + (t[[j, j], j]) \circ \gamma_{3n}(\alpha).$$

This implies the proposition $\Gamma_n^r \subset \Lambda_n^r$. We consider the diagram

(2) 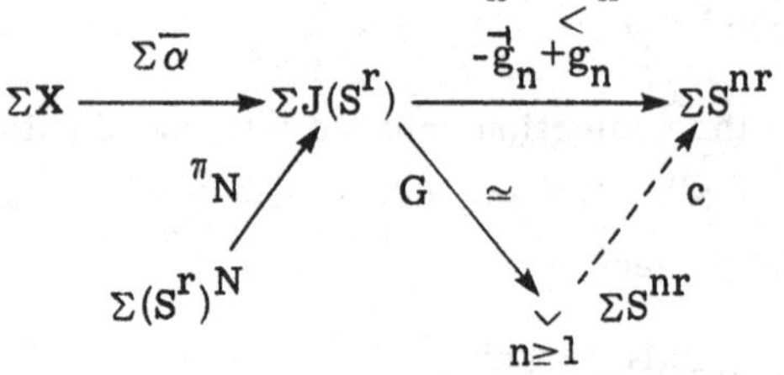

where $N > \dim X$. By definition (0.11) we have

$$(3)\qquad -\overset{\dashv}{\gamma}_n(\alpha) \overset{<}{+} \gamma_n(\alpha) = (-\bar{g}_n \overset{<}{+} g_n) \circ \Sigma\bar{\alpha}$$

In the diagram

$$(4)\qquad G = \sum_{n\geq 1} i_n \circ \bar{g}_n$$

is a homotopy equivalence. Here $\bar{g}_n = \gamma_n(\bar{g}) : \Sigma J(S^r) \to \Sigma S^{nr}$ is the James-Hopf invariant of the adjoint $\bar{g} : \Sigma J(S^r) \to \Sigma S^r$ of g in II (2.1). Moreover i_n is the inclusion into the wedge. Since G is a homotopy equivalence there exists a factorisation $c = (c_n, n \geq 1)$ with

$$(5)\qquad -\bar{g}_n \overset{<}{+} g_n \simeq c \circ G.$$

We therefore know by (4) and (3)

$$(6)\qquad -\overset{\dashv}{\gamma}_n(\alpha) + \overset{<}{\gamma}_n(\alpha) = \sum_{n\geq 1} c_n \circ \gamma_n(\alpha) .$$

For the computation of c_n we use (2. 10). By (2. 8) and (2. 5) we know

$$(7)\qquad \begin{aligned} (-\overset{\dashv}{g}_n + \overset{<}{g}_n) \circ \Sigma\pi_N &= - \overset{\dashv}{\sum_{a\subset \overline{N}}} p_a + \overset{<}{\sum_{a\subset \overline{N}}} p_a + \\ &+ \sum_P (p_a, p_{a'}) + \sum_Q ((p_a , p_{a'}), p_{a''}). \end{aligned}$$

The sums over P and Q can be taken in arbitrary order. We choose such an ordering of the sums so that they factor over

$$(8)\qquad G \circ \Sigma\pi_N = \overset{\text{lex}}{\sum_a} i_{\#a} p_a .$$

This is possible since $<$ and $\dashv$ are <u>admissible</u> orderings. For example we have by (2. 9)

$$\sum_P (p_a, p_{a'}) = \overset{\text{lex}}{\sum_{\substack{b\subset\overline{N}\\ \#b=2n}}} c_{2n} \circ p_b$$

where $c_{2n} = s[j, j]$ with

$$(9)\qquad s = \sum (\varepsilon_{a, b})^n .$$

The sum is taken over all partitions (a, b) of $\overline{2n}$ with $\#a = \#b = n$ and $a < b$ and $b \dashv a$.

Because of (7) only c_{2n} and c_{3n} are non trivial in (6) and this proves (2). //

§ 3. Deviation from commutativity of spherical cup products

(3. 1) Theorem. *Let X and Y be finite dimensional. For $\alpha \in [\Sigma X, \Sigma S^n]$ and $\beta \in [\Sigma Y, \Sigma S^m]$ we have*

$$\alpha \# \beta = \alpha \underline{\#} \beta + ((-1)^{nm+m}[j, j])(\gamma_2(\alpha) \# \gamma_2(\beta))$$

where $j = j_{n+m}$. *If* $X = Y$ *we get*

$$\alpha \cup \beta = ((-1)^{nm}j)(\beta \cup \alpha) + ((-1)^{nm+m}[j, j])(\gamma_2(\alpha) \cup \gamma_2(\beta)).$$

Proof. We choose N with $\dim X < N$ and $\dim Y < N$. Let

$$M = \{ij \mid 1 \le i,\ j \le N\}.$$

We define two orderings on M

$$ij < rs \Leftrightarrow i < r \quad \text{or} \quad i = r \text{ and } j < s$$

$$ij - rs \Leftrightarrow j < s \quad \text{or} \quad j = s \text{ and } i < r$$

For the adjoint

$$f_i = \bar{g} : \Sigma J(S^{n_i}) \to \Sigma S^{n_i}$$

$(n_1 = n,\ n_2 = m)$ of g in II (2.1) we have

$$\alpha = f_1(\Sigma\bar{\alpha}),$$
$$\beta = f_2(\Sigma\bar{\beta}),$$

For the differences

$$F' = -(\alpha \underline{\#} \beta) + (\alpha \# \beta),$$
$$F = -(f_1 \underline{\#} f_2) + (f_1 \# f_2),$$

we thus obtain the homotopy commutative diagram

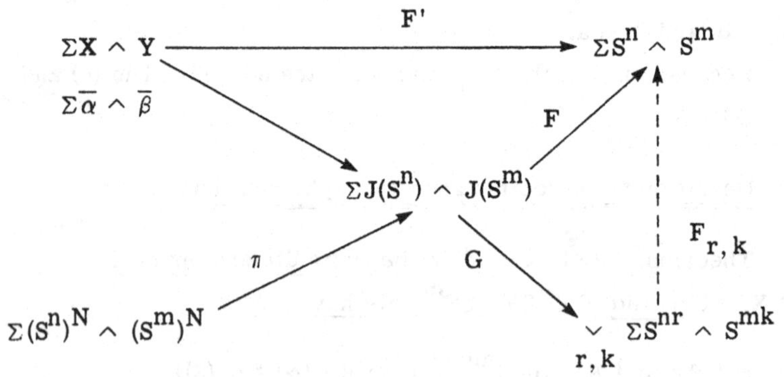

π is defined in the obvious way as in (2.6) and G is a homotopy equivalence given by

$$G = \sum_{r,k} j_{r,k} \circ (\bar{g}_r \# \bar{g}_k)$$

with $\bar{g}_r = \gamma_r(\bar{g})$ as in the proof of (1. 4). We want to compute the factorization $F_{r,k}$, $r, k > 1$.

From the definition of $\#$ and $\underline{\#}$ we obtain the formula (compare (2. 8) with $r = 1$ and $g_1 = g$):

$$F\pi = -(\overset{-}{\sum_{ij}} p_i \# q_j) + (\overset{<}{\sum_{ij}} p_i \# q_j)$$

with the orderings on M as defined above and where

$$\text{and} \quad \begin{array}{l} p_i : \Sigma(S^n)^N \to \Sigma S^n \\ q_i : \Sigma(S^m)^N \to \Sigma S^m \end{array}$$

are the projections onto the i-th coordinate. From (2. 5) we get

$$F\pi = \sum_P [ij, i'j'] + \sum_Q [[ij, i'j'], i"j"]$$

where ij stands for $p_i \# q_j$. Since the indices ij, i'j', i"j" have to be pairwise disjoint we have

$$P = \{(ij, i'j') | i < i' \text{ and } j > j'\}$$

$$Q = \left\{ (ij, i'j', i"j") \;\middle|\; \begin{array}{l} i < i' \text{ and } j > j' \text{ such that} \\ (j < j") \text{ or } (j > j" \text{ and } i' < i"). \end{array} \right\}$$

We order P lexicographically so that the sum $\sum_P$ factors over $\bar{g}_2 \# \bar{g}_2$. Thus we see that

$$F_{22} = (-1)^{nm+m^2} [j_{n+m}, j_{n+m}].$$

We have the sign $(-1)^{m^2}$ since $j > j'$. When we have shown that the sum $\sum_Q$ is trivial, (3. 1) is proved by use of (2. 10). Look at

$$Q^{III}_{c,c'} = \{(ij, i'j', i''j'') \in Q \mid c = \{i, i', i''\}, c' = \{j, j', j''\}\}$$

for given sets $c, c' \subset \{1, \dots, N\}$ with $\#c = \#c' = 3$. $Q_{c,c'}$ has 5 elements, and the corresponding summands cancel. For example: with $c = \{1, 2, 3\}$ and $c' = \{1, 2, 3\}$ we get

$Q_{c,c'}$	n even m odd	n odd m even
12, 21, 33	-1	+1
13, 21, 32	1	+1
13, 22, 31	-1	+1
12, 31, 23	-1	-1
22, 31, 13	-1	+1

Since $[[j, j], j]$ is at most of order 3, $Q_{c,c'}$ vanishes. //

§ 4. Cup products of spherical Hopf invariants

Let $\mathbb{N} = \{1, 2, \dots\}$ be the set of natural numbers. A word $A = a^1 \dots a^r$, $r \geq 1$, of pairwise disjoint non empty subsets $a^i \subset \mathbb{N}$ is a partition of $\underline{A} = a^1 \cup \dots \cup a^r$. For a subset $a \subset \mathbb{N}$ we have $a = \{a_1 < \dots < a_k\}$ where $k = \#a$ is the number of elements of a. Each partition $A = a^1 \dots a^r$ with $\underline{A} = \bar{n} = \{1, \dots, n\}$ determines the shuffle permutation $\sigma(A) \in S_n$ with

$$\sigma(A)(k_1 + \dots + k_{i-1} + j) = (a^i)_j$$

for $k_i = \#a^i$ $(i = 1, \dots, r)$ and $1 \leq j \leq k_i$. The shuffle sign $\varepsilon(A)$ is the sign of this permutation $\sigma(A)$.

Let Par be the set of all partitions A in $\mathbb{N}$.

Lemma (2.5) gives rise to the following definition:

(4.1) **Definition.** Let M be a subset of Par such that for all $A \in M$ we have $\#\underline{A} = n$, and let $<$, $\dashv$ be two orderings on M. We associate with $(M, <, \dashv)$ and $k \in \mathbb{N}$ three integers N^k, P^k and Q^k:

$$N^k(M) = \sum_{\substack{A \in M \\ \underline{A}=\overline{n}}} \varepsilon(A)^k$$

$$P^k(M, <, \dashv) = \sum_{(A, B) \in P} \varepsilon(AB)^k$$

where P is the set of all pairs (A, B) with $A, B \in M$, $\underline{A} \cap \underline{B} = \emptyset$, $\underline{A} \cup \underline{B} = \overline{2n}$ and $A < B$, $B \dashv A$. Furthermore let

$$Q^k(M, <, \dashv) = \sum_{(A, B, C) \in Q} \varepsilon(ABC)^k$$

where Q is the set of all triples (A, B, C) with $A, B, C \in M$, $\{\underline{A}, \underline{B}, \underline{C}\}$ pairwise disjoint and $\underline{A} \cup \underline{B} \cup \underline{C} = \overline{3n}$ subject to the condition $(A < B$ and $B \dashv A)$ and $((B < C$ and $C \dashv A)$ or $(A \dashv C))$. For the definition of P^k and Q^k we need to know the ordering $A < B$ or $A \dashv B$ only if $\underline{A} \cap \underline{B} = \emptyset$.

(4. 2) **Proposition.** Let X be finite dimensional and let $\alpha \in [\Sigma X, \Sigma S^k]$. In $[\Sigma X, \Sigma S^{(n+m)k}]$ we have the equation $(n, m \geq 1)$:

$$\begin{aligned} \gamma_n(\alpha) \cup \gamma_m(\alpha) = {} & (N^k j)\, \gamma_{n+m}(\alpha) + \\ & (P^k[j, j])\, \gamma_{2(n+m)}(\alpha) + \\ & (Q^k[[j, j], j])\, \gamma_{3(n+m)}(\alpha) \end{aligned}$$

N^k, P^k and Q^k are given by the set

$$M_{n, m} = \{ab \in \mathrm{Par} \mid \#a = n,\ \#b = m\}$$

with the orderings

$$ab < a'b' \Leftrightarrow \mathrm{Min}(a) < \mathrm{Min}(a')$$
$$a'b' \dashv ab \Leftrightarrow \mathrm{Min}(a \cup b) < \mathrm{Min}(a' \cup b')$$

For the number

$$N^k_{n, m} = N^k(M_{n, m}) = \sum_{\substack{a \cup b = \overline{n+m} \\ \#a = n \\ \#b = m}} \varepsilon(ab)^k$$

defined in (4.1) we know for $n, m, k \geq 1$

$$(4.3)\quad N^k_{n,m} = \begin{cases} \binom{n+m}{n} & , \ k \text{ even} \\ 0 & , \ nmk \text{ odd} \\ \binom{[(n+m)/2]}{[n/2]} & , \ \text{otherwise.} \end{cases}$$

where $[n/2] = n'$ if $n = 2n'$ or $n = 2n' + 1$.

Proof of (4.2). Let $\bar{\alpha} : X \to J(S^k)$ be the adjoint of α. Since

$$\gamma_n(\alpha) = \bar{g}_n \circ (\Sigma\bar{\alpha})$$

we get

$$\gamma_n(\alpha) \cup \gamma_m(\alpha) = (\bar{g}_n \cup \bar{g}_m) \circ (\Sigma\bar{\alpha}) .$$

We want to compute the difference of the mappings

$$\Sigma X \xrightarrow{\Sigma\bar{\alpha}} \Sigma J(S^k) \underset{(N^k j)\bar{g}_{n+m}}{\overset{\bar{g}_n \cup \bar{g}_m}{\rightrightarrows}} \Sigma S^{(n+m)k}$$

We choose N with $\dim X < N$ and we set

$$M = \{ab \in M_{n,m} \mid a \cup b \subset \bar{N}\}$$

We introduce two orderings $<$ and $\dashv$ on M, namely

$$ab < a'b' \Leftrightarrow a \underset{\text{lex}}{<} a' \text{ or } a = a' \text{ and } b \underset{\text{lex}}{<} b'$$

$$ab \dashv a'b' \Leftrightarrow a \cup b \underset{\text{lex}}{<} a' \cup b' \text{ or}$$

$$a \cup b = a' \cup b' \text{ and } ab < a'b'$$

where we use the lexicographical ordering from the left $\underset{\text{lex}}{<}$. From (2.8) and (0.5) we know

$$f_1 = (\bar{g}_n \cup \bar{g}_m)(\Sigma\pi_N) = \sum_{ab \in M}^{<} p_a \cup p_b .$$

On the other hand we have

$$f_2 = (N^k(M_{n,m})\, j)\bar{g}_{n+m}(\Sigma\pi_N) = \overset{\dashv}{\sum_{ab\in M}} p_a \cup p_b$$

where we use $p_a \cup p_b = (\varepsilon^k_{ab}\, j)\; p_{a\cup b}$, see (2. 9).

We now can apply lemma (2. 5), and using (2. 10) obtain proposition (4. 2). //

(4. 4) **Lemma.** The number $P^k_{n,m} = P^k(M_{n,m}, <, \dashv)$ in (4. 2) is given by the formula

$$P^k_{n,m} = (-1)^{nmk+mk} N^k_{n-1,n} N^k_{m,m-1} \cdot N^k_{2n,2m-1}\,.$$

Proof. $P^k_{n,m}$ is the sum of all elements

$$\varepsilon(a\ b\ c\ d)^k \in \{1, -1\}$$

with $a\ b\ c\ d = \overline{2n+2m}$, $\#a = \#c = n$, $\#b = \#d = m$, such that $a_1 < c_1$ and $\mathrm{Min}(c \cup d) < \mathrm{Min}(d \cup b)$. This implies $d_1 = 1$. Therefore

$$P^k_{n,m} = \sum_{a_1 < c_1} \varepsilon(a,\ b,\ c,\ 1,\ d')^k$$

with $1 \cup d' = d$. Since

$$\varepsilon(a,\ b,\ c,\ 1,\ d') = (-1)^{nm}\varepsilon(a,\ c)\,\varepsilon(a \cup c,\ b,\ 1,\ d')$$

we get

$$P^k_{n,m} = (-1)^{nmk} \sum_{ub1d'} \varepsilon(a,\ b,\ 1,\ d')^k \cdot \sum_{\substack{a\cup b = u \\ u_1 = a_1 < c_1}} \varepsilon(a,\ c)^k$$

$$= (-1)^{nmk} \sum \varepsilon(a,\ b,\ 1,\ d')^k \cdot N^k_{n-1,n}$$

$$= (-1)^{nmk}(-1)^{mk} N^k_{n-1,n} N^k_{m,m-1} N^k_{2n,2m-1}. \quad //$$

We are interested in the following special case of (4. 2), see § 2.

(4. 5) **Corollary.** Let $\alpha \in [\Sigma X, \Sigma S^k]$ where k is odd, then for $n \geq 1$

$$\gamma_{2n+1}(\alpha) = \gamma_{2n}(\alpha) \cup \alpha + \binom{2n-1}{n} [j,\ j]\, \gamma_{2(2n+1)}(\alpha)$$

Proof. We consider $\gamma_{2n}(\alpha) \cup \gamma_1(\alpha)$ in (4. 2), then $N^k = 1$ by (4. 3) and $P^k = -\frac{1}{2}\binom{2n}{n}$ by (4. 4). Thus the proposition follows from

(1) $\qquad Q^k = Q^k(M_{2n,\,1}, <, \dashv\,) \equiv 0 \mod 3.$

Q^k is the sum of all elements

$$\varepsilon(ab\ cd\ fg) \in \{1, -1\}$$

with $\underline{a\,b\,c\,d\,f\,g} = \overline{6n+3}$, $\#a = \#c = \#f = 2n$ and $\#b = \#d = \#g = 1$ such that

(2) $\qquad a_1 < c_1$ and $\mathrm{Min}(c \cup d) < \mathrm{Min}(a \cup b)$

and

(3) $\qquad c_1 < f_1$ and $\mathrm{Min}(f \cup g) < \mathrm{Min}(a \cup b)$

(4) $\qquad$ or $\mathrm{Min}(a \cup b) < \mathrm{Min}(f \cup g)$.

(2) and (4) imply $d = 1$ and $a_1 = 2$ or $b = 2$. On the other hand (2) and (3) imply $(d, g) = (1, 2)$ or $(d, g) = (2, 1)$.

We now fix disjoint subsets $x, y, z \subset \overline{6n+3}$ with $\#x = \#y = \#z = 2n$ and $x_1 < y_1 < z_1$. The subsum of Q^k with all indices $a, b, c \in \{x, y, z\}$ contains exactly the following summands ε where $\tau = \varepsilon(x, y, z)$.

	a	b	c	d	f	g	ε
(5)	x	2	y	1	z	g	$-\tau$
	x	2	z	1	y	g	$-\tau$
	y	z	z	1	x	g	$-\tau$
(6)	x	b	y	1	z	g	τ
	x	g	y	1	z	b	$-\tau$
	x	b	z	1	y	g	τ
	x	g	z	1	y	b	$-\tau$
(7)	x	b	y	1	z	2	τ
	x	b	y	2	z	1	$-\tau$

(5) and (7) are only summands in case $1, 2 \notin \underline{x\ y\ z}$ and (6) describes summands only if $x_1 = 2$, we may assume $b < g$ in (6).

Since all these partial sums are congruent 0 mod 3, we have proved (1). //

By (4. 5) Hopf invariants of even degree determine all Hopf invariants. Therefore it is not necessary to compute Q^k in (4. 2), since for even degree $(n + m)k$ the triple product vanishes, see (1. 1). For computations with formula (4. 2) it is enough to know (4. 4).

We still consider the following special case of (4. 2):

(4. 6) **Corollary.** Let X be finite dimensional and $\alpha \in [\Sigma X, \Sigma S^k]$ then

$$\alpha \cup \alpha = \begin{cases} 0 & \text{if } k \text{ is odd} \\ (2j)\,\gamma_2(\alpha) & \text{if } k \text{ is even.} \end{cases}$$

Proof. Consider $\gamma_1(\alpha) \cup \gamma_1(\alpha)$ in (4. 2). If k is odd we know $N^k_{1,1} = 0$. Thus

$$\begin{aligned} \alpha \cup \alpha = {} & P^k_{1,1}\,[j, j]\,\gamma_4(\alpha) \\ & + Q^k(M_{1,1})[[j, j], j]\,\gamma_6(\alpha)\,. \end{aligned}$$

From the definition of P^k and Q^k we get as in the proof of (4. 4)

$$\begin{aligned} P^k_{1,1} & = \varepsilon(2\ 3\ 4\ 1) + \varepsilon(2\ 4\ 3\ 1) + \varepsilon(3\ 2\ 4\ 1) \\ & = +\,1, \end{aligned}$$

and we get $Q^k(M_{1,1}) = 0$. Therefore the proposition follows from (6. 2). If k is even we have by (4. 4) and (4. 2)

$$\alpha \cup \alpha = (2j)\,\gamma_2(\alpha) + 3[j, j]\,\gamma_4(\alpha)$$

Again the proposition follows from (6. 2). //

§ 5. Hopf invariants of a Hopf invariant, of a sum and of a cup product

With (4. 3) we define the number

(5.1) $$M^k_{n,m} = \prod_{i=1}^{n-1} N^k_{m-1,im}$$

Then for k even we have

$$M^k_{n,m} = (n \cdot m)!/((m!)^n n!)$$

(5.2) Theorem. Let X be finite dimensional and $\alpha \in [\Sigma X, \Sigma S^k]$. In $[\Sigma X, \Sigma S^{nmk}]$ we have the equation $(n, m \geq 1)$

$$\gamma_n \gamma_m(\alpha) = (M^k_{n,m} j)\gamma_{nm}(\alpha).$$

Proof. Along the same lines as in the proof of (4.2) we obtain the formula

$$\begin{aligned}\gamma_n\gamma_m(\alpha) &= (N^k j)\gamma_{nm}(\alpha)\\ &+ (P^k[j, j])\gamma_{2nm}(\alpha)\\ &+ (Q^k[[j, j], j])\gamma_{3nm}(\alpha)\end{aligned}$$

where N^k, P^k, Q^k are given by the set

$$M^n_m = \left\{ A = a^1 \ldots a^n \in \mathrm{Par} \;\middle|\; \begin{array}{l} \#a^1 = \ldots = \#a^n = m \\ \mathrm{Min}\, a^1 < \ldots < \mathrm{Min}\, a^n \end{array} \right\}$$

with the orderings

$$A < A' \Leftrightarrow \mathrm{Min}(a^1) < \mathrm{Min}(a'^1)$$

$$A \dashv A' \Leftrightarrow \mathrm{Min}(\underline{A}) < \mathrm{Min}(\underline{A}')$$

Since $\mathrm{Min}(a^1) = \mathrm{Min}(\underline{A})$ we see $A < A' \Leftrightarrow A \dashv A'$. Therefore there are no pairs (A, A') with $A < A'$ and $A' \dashv A$ and this implies $P^k = Q^k = 0$.

Moreover we can prove that in fact the number $N^k(M^n_m)$ defined in (4.1) is equal to $M^k_{n,m}$ defined in (5.1). //

In II § 2 we have already seen:

(5.3) Proposition. Let $\alpha, \beta \in [\Sigma X, \Sigma S^k]$ then

$$\gamma_n(\alpha + \beta) = \gamma_n\alpha + \left(\sum_{i=1}^{n-1} \gamma_{n-i}(\alpha) \cup \gamma_i(\beta)\right) + \gamma_n(\beta)\,.$$

(5.4) **Proposition.** Let $\alpha \in [\Sigma X, \Sigma S^k]$, $\beta \in [\Sigma X, \Sigma S^r]$, $k, r, n \geq 1$, then

$$\gamma_n(\alpha \cup \beta) = ((-1)^{kr\cdot\binom{n}{2}} j)((\gamma_n\alpha) \cup \beta^n)$$

where $\beta^n = \beta \cup \ldots \cup \beta$ is the n-fold cup product.

(5.2) and (5.3) are special properties of the lexicographical ordering from the left.

§ 6. Hopf invariants of a composition element

In this section we exhibit an expansion formula for the Hopf invariant $\gamma_n(\beta \circ \alpha)$ of a composition element $\beta \circ \alpha$, where β maps into a sphere. This yields for the suspended Hopf invariants $\lambda_n = \Sigma^{n-1}\gamma_n$, the expansion of J. M. Boardman and B. Steer, see 3.16 of [13]. For $\gamma_2(\beta \circ \alpha)$ we improve a formula of B. Steer in [39].

(6.1) **Proposition.** Let X and A be connected finite dimensional spaces and let A be a Co-H-space. For the composition

$$\Sigma X \xrightarrow{\alpha} \Sigma A \xrightarrow{\beta} \Sigma S^k$$

we obtain the Hopf invariant in $[\Sigma X, \Sigma S^{2k}]$ by

$$\gamma_2(\beta\alpha) = (\gamma_2\beta)\alpha + (\beta \# \beta)(\gamma_2\alpha) +$$
$$[j, j] \circ \{(\beta \# \beta \# (\gamma_2\beta))(\gamma_3\alpha) + (\beta \# \beta \# \beta \# \beta)(\gamma_4\alpha)\}$$

(6.2) **Corollary.** Let X be connected and finite dimensional. For $\alpha \in [\Sigma X, \Sigma S^k]$ with $k \geq 1$ we have in $[\Sigma X, \Sigma S^{2k}]$

$$[j, j]\gamma_4(\alpha) = 0.$$

Proof. Take $A = S^k$ and $\beta = j$ in (6.1). //

(6.3) **Corollary.** Let X be a Co-H-space. Then the composition

$$\Sigma X \xrightarrow{\alpha} \Sigma S^n \xrightarrow{\beta} \Sigma S^k$$

has the Hopf invariant

$$\gamma_2(\beta\alpha) = (\gamma_2\beta)\alpha + (\beta \# \beta)\gamma_2\alpha .$$

Proof of (6.3). For $\beta : \Sigma S^n \to \Sigma S^k$ we have by (6.2), (1.4)

$$[j_{2k}, j_{2k}](\beta \# \beta \# \beta \# \beta)\gamma_4\alpha = [\beta \# \beta, \beta \# \beta]\gamma_4\alpha$$
$$= (\beta \# \beta)[j_{2n}, j_{2n}]\gamma_4\alpha = 0.$$

Moreover from (4.5) it follows that the term in (6.1) involving $\gamma_3(\alpha)$ is trivial if n is odd. If n is even, $\gamma_3\alpha$ has order at most 3, see chapter IV. Since [j, j] has order ≤ 2 the term involving $\gamma_3\alpha$ also vanishes if n is even. //

Remark. In [39] B. Steer calculated the Hopf invariant of a composition element $\beta \circ \alpha \in \pi_{r+1}(S^{k+1})$. He obtains in 4.7 of [39] a formula similar to (6.3) but with a correction term Δ. (6.3) settles in the affirmative his question whether this correction term is trivial. The quite intricate methods of [39] are different from ours and yield the correction term Δ in a different form.

We obtain (6.1) as a special case of the following more general result.

(6.4) **Proposition.** Let X and A be finite dimensional connected spaces. For the James-Hopf invariant of the composition

$$\Sigma X \xrightarrow{\alpha} \Sigma A \xrightarrow{\beta} \Sigma S^k$$

we obtain

$$\gamma_n(\beta \circ \alpha) = \left(\sum_{r=1}^{\infty} \Gamma_n^r(\beta) \circ \gamma_r(\alpha) \right) + \Delta$$

where

$$\Gamma_n^r(\beta) = \sum_{i_r=1}^{n} \sum_{i_{r-1}=1}^{n-i_r} \cdots \sum_{i_2=1}^{n-i_r-\dots-i_3} \gamma_{i_1}(\beta) \# \dots \# \gamma_{i_r}(\beta)$$

with $i_1 = n - i_r - \dots - i_2$. Δ vanishes under suspension. If nk is even, Δ is at most an element of order 2 which we define as follows. The case nk odd is settled by (4.5).

In (0.8) we define the lexicographical ordering $<_{lex}$ of subsets of $\mathbb{N}$. For $i, i' \in \mathbb{Z}^N = \mathbb{Z} \times \dots \times \mathbb{Z}$ we have correspondingly the lexicographical ordering

$$i \lhd_{lex} i' \Leftrightarrow \begin{cases} \text{For } j \text{ with } i_\alpha = i'_\alpha \\ (\alpha < j \le N) \text{ and } i_j \ne i'_j \\ \text{let } i_j < i'_j . \end{cases}$$

Now we choose N with $N > \dim X$. For the subset of $\mathbb{Z}^N$

$$M_n = \{(i_1, i_2, \dots, i_N) \mid i_\alpha \ge 0, \ \sum_{\alpha=1}^{N} i_\alpha = n\}$$

we define two orderings

$$i < i' \Leftrightarrow (i_N, \dots, i_2) \lhd_{lex} (i'_N, \dots, i'_2)$$

$$i \dashv i' \Leftrightarrow \begin{cases} i_+ \lhd_{lex} i'_+ \text{ and} \\ \text{if } i_+ = i'_+ \text{ then } i < i' \end{cases}$$

where

$$i_+ = \{\alpha \in \{1, \dots, N\} \mid i_\alpha \ne 0\}.$$

As in (3.1) we associate with $(M_n, <, \dashv)$ the set

$P \subset M \times M$, with

$(i, i') \in P \Leftrightarrow i < i'$ and $i' \dashv i$.

For $i \in M$ we set

$$\gamma_i = \gamma_{i_1}(\beta) \# \dots \# \gamma_{i_N}(\beta)$$

where we omit the factors $\gamma_{i_\alpha}(\beta)$ with $i_\alpha = 0$. With this notation we obtain Δ in the proposition by

(6.5) $\Delta = \sum_{m\geq 1} P_m \circ \gamma_m(\alpha)$ for nk even

where

$$P_m = \sum_{\substack{(a,b)\epsilon P_- \\ a_+ \cup b_+ = \bar{m}}} [\gamma_a, \gamma_b] \circ T(a_+, b_+)$$

For subsets a^i of $\bar{m} = \{1, 2, \dots, m\}$ with $a^1 \cup \dots \cup a^k = \bar{m}$ let $T(a^1, \dots, a^k)$ be the suspension of

$$T : A^{\wedge m} \to A^{\wedge \# a^1} \wedge \dots \wedge A^{\wedge \# a^k}$$

mapping $x_1 \wedge \dots \wedge x_m$ to $x(a^1) \wedge \dots \wedge x(a^k)$ where $x(a) = x_{a_1} \wedge \dots \wedge x_{a_r}$ for $a = \{a_1 < \dots < a_r\}$. If A is a Co-H-space T is null homotopic if $a^1, \dots, a^k$ are not pairwise disjoint.

Proof of (6.1). The result in (6.4) gives us the formula

$$\gamma_2(\beta \circ \alpha) = (\gamma_2\beta)\alpha + (\beta \# \beta) \circ \gamma_2 \alpha + \Delta \qquad \text{where}$$

$$\Delta = [\beta \# \beta, \gamma_2\beta] \circ (\gamma_3 \alpha) + [\beta \# \beta, \beta \# \beta] \circ T_{2314} \circ (\gamma_4 \alpha)$$

from which we derive the formula in (6.1). In fact M_2 contains only two types of elements, namely

$$\begin{Bmatrix} 0 \dots 010 \dots 010 \dots 0 \\ 0 \dots 020 \dots 0 \end{Bmatrix} = M_2 .$$

Now we see P_2 is trivial. P_3 has only one summand with the index pair (a, b) with a = 110 and b = 002. Moreover, P_4 has only one summand with the index pair (a, b) with a = 0110 and b = 1001. //

Proof of (6.4). For the adjoints $\bar{\alpha}$, $\bar{\beta}$ (see (0.9)) we have

(1) $\overline{\beta \circ \alpha} = \bar{\beta}_\infty \circ \bar{\alpha}$

where $\overline{\beta}_\infty = JA \to JS^k$ is the homomorphic extension of $\overline{\beta}$. We now consider the diagram

$$\begin{array}{ccccc}
\Sigma A^N & & \xrightarrow{\quad \Sigma\overline{\beta}^N \quad} & & \Sigma(JS^k)^N \\
\Sigma\pi_N \Big\downarrow & & (2) & & \Big\downarrow \Sigma\mu \\
\Sigma JA & & \xrightarrow{\quad \Sigma\overline{\beta}_\infty \quad} & & \Sigma JS^k \\
\simeq \Big\downarrow G & & (3) & & \Big\downarrow \overline{g}_n = \gamma_n(\overline{g}) \\
\bigvee_{i\geq 1} \Sigma A^{\wedge i} & & \xrightarrow[\quad \Gamma_n^*(\beta) \quad]{} & & \Sigma S^{kn}
\end{array}$$

$A^N = A \times \ldots \times A$ is the N-fold product and μ is the multiplication. π_N is defined as in (2.6). Thus (2) commutes.

The homotopy equivalence G is given by

(4) $$G = \sum_{n\geq 1} i_n \overline{g}_n$$

where i_n is the inclusion of $\Sigma A^{\wedge n}$. We want to compute the difference of the mappings.

(5) $$f_1 = \overline{g}_n(\Sigma\overline{\beta}_\infty)$$
$$f_2 = \Gamma_n^*(\beta) \circ G$$

in (3). Clearly because of (1) and (4) we know for Δ in the proposition

(6) $$\Delta = (\Sigma\overline{\alpha})\,(-f_2 + f_1).$$

As in II (2.15) we see

(7) $$\overline{g}_n(\Sigma\mu) = \sum_{i_N=0}^{n} \; \sum_{i_{N-1}=0}^{n-i_N} \cdots \sum_{i_2=0}^{n-i_N-\ldots-i_3} \overline{g}_{i_1} q_1 \cup \ldots \cup \overline{g}_{i_N} q_N$$

where $i_1 = n - i_N - \ldots - i_2$ and where $q_i : X^N \to X$ is the projection onto the i-th coordinate. We omit the factor $g_{i_\alpha} q_\alpha$ if $i_\alpha = 0$.

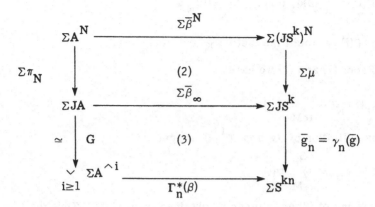

III

For $i = (i_1, \ldots, i_N) \in M$ we set

$$\Gamma(i) = \gamma_{i_1}(\beta)q_1 \cup \ldots \cup \gamma_{i_N}(\beta)q_N .$$

Again in $\Gamma(i)$ we omit the factor $\gamma_{i_\alpha} q_\alpha$ if $i_\alpha = 0$.

From (7) and (2) we have

$$(8) \qquad f_1(\Sigma\pi_N) = \overset{<}{\sum_{i \in M}} \Gamma(i) .$$

From the definition of G and $\Gamma^i_n(\beta)$ we get

$$(9) \qquad f_2(\Sigma\pi_N) = \overset{\dashv}{\sum_{i \in M}} \Gamma(i)$$

The difference of (8) and (9) can be obtained from (2. 5). Thus the proposition is proved. //

As an example we know for the Whitehead product $[j_k, j_k] \in \pi_{2k+1}(S^{k+1})$

$$\begin{cases} \gamma_2([j_k, j_k]) = 2j & \text{if } k \text{ is odd} \\ \gamma_n([j_k, j_k]) = 0 & \text{otherwise } n > 1. \end{cases}$$

Since $\Sigma[j_k, j_k] = 0$ we get

(6. 6) **Corollary.** For $\alpha : \Sigma X \to \Sigma S^{2k}$ we have

$$\gamma_m([j_k, j_k]\alpha) = \begin{cases} (2^n j)(\gamma_n \alpha) & \text{if } k \text{ is odd and } m = 2n \\ 0 & \text{otherwise.} \end{cases}$$

IV. HIGHER ORDER HOPF INVARIANTS ON SPHERES

§1. Examples of higher order Hopf invariants on spheres

Again we consider the generalized homotopy groups of spheres $[\Sigma X, \Sigma S^k]$ where ΣX is the suspension of a finite dimensional connected polyhedron X. On such groups the James Hopf invariants

(1.1) $\gamma_n : [\Sigma X, \Sigma S^k] \to [\Sigma X, \Sigma S^{nk}]$

are defined. In general these are not homomorphisms of groups. They are non trivial as is easily seen by

(1.2) **Proposition.** If $x \in H^k(X, \mathbb{Q})$ has non trivial cup product power x^n there is a map $f : \Sigma X \to \Sigma S^k$ with rational degree x such that $\gamma_n(f)$ has rational degree $x^n/n!$

However, if X is a Co-H-space, all cup products vanish. In this case most of the invariants γ_n vanish too. We prove in §2:

(1.3) **Theorem.** Let X be a finite dimensional Co-H-space. Then for $n \geq 2$ all possible definitions of James-Hopf invariants yield the same function (1.1) which is a homomorphism of abelian groups. Moreover, if k is even we have

$p\gamma_n(\alpha) = 0$ if $n = p^v$, $v \geq 1$, p a prime,

$\gamma_n(\alpha) = 0$ for other $n \geq 2$.

If k is odd we have

$p\gamma_n(\alpha) = 0$ if $n = 2p^v$, $v \geq 1$, p a prime,

$\gamma_n(\alpha) = 0$ for other $n \geq 3$.

This improves results of M. Barratt [6] and of J. Boardman,

B. Steer, see 3.18 in [13]. The theorem shows that the only Hopf invariant for a Co-H-space X, which might be of infinite order, is the classical one:

(1.4) $\gamma_2 : [\Sigma X, \Sigma S^k] \to [\Sigma X, \Sigma S^{2k}]$, k odd.

In fact, as we know by the result of Adams for $X = S^{2k}$ the image of γ_2 is $[\Sigma S^{2k}, \Sigma S^{2k}] = \mathbb{Z}$ if $k = 1, 3, 7$ and is $2\mathbb{Z}$ for other odd k.

We now exhibit examples of Co-H-spaces X, namely spheres and Moore spaces, with non trivial higher order invariants (1.1) $n \geq 3$. To my knowledge these are the first examples to be described in the literature. The examples show that theorem (1.3) to some extent is best possible.

Let

(1.5) $P^{k+1}(p) = S^k \cup_p e^{k+1}$

be the Moore space for $\mathbb{Z}_p$ in dimension k. Clearly for $k \geq 2$ $P^{k+1}(p) = \Sigma P^k(p)$ is a Co-H-space. We have the pinch map

$$\mu : P^{k+1}(p) \to S^{k+1}.$$

For a prime p we consider an element $\alpha_p \in \pi_{2p}(S^3)$ which generates the p-primary component. Since α_p has order p we have an extension

$$\underline{\alpha}_p \in [\Sigma P^{2p}(p), \Sigma S^2]$$

of α_p. We prove

(1.6) Theorem. $\gamma_p(\underline{\alpha}_p) \in [\Sigma P^{2p}(p), \Sigma S^{2p}]$ is an element of order p. In fact $\gamma_p(\underline{\alpha}_p)$ generates the same sub-group as the pinch map $\Sigma\mu$ (p an odd prime).

It is well known that α_p is a Co-H-map. Clearly $\underline{\alpha}_p$ is no Co-H-map.

Proof. The reduced product $J(S^2)$ is a CW-complex with the n-fold reduced product $J_n(S^2)$ as a 2n-skeleton. Let

$$[i_2]^p : S^{2p-1} \to J_{p-1}(S^2)$$

be the attaching map of the 2p cell $e_{2p} = J_p S^2 - J_{p-1} S^2$. In [2] we prove that $[i_2]^p$ is divisible by p. So let α be given with

(*) $\quad p\alpha = [i_2]^p$.

Then the composition, with $\bar{r}$ = identity on JS^2,

$$S^{2p} \xrightarrow{\Sigma\alpha} \Sigma J_{p-1}(S^2) \hookrightarrow \Sigma JS^2 \xrightarrow{r} \Sigma S^2$$

generates the p-primary component of $\pi_{2p}(S^3)$, see [8]. Because of (*) α extends to

$$\underline{\alpha} : P^{2p}(p) \to J(S^2)$$

and $\underline{\alpha}$ is of degree 1 on the 2p cell. Therefore the composition $g_p \underline{\alpha}$ is

$$P^{2p}(p) \xrightarrow[\mu]{} S^{2p} \hookrightarrow J(S^{2p}) . \quad //$$

In [1] D. W. Anderson proves the following formula which relates Hopf invariants with the e-invariant of complex K-theory. Let k be odd and $2t + k + 1 > 2r \cdot t + 1$. Then for $\alpha \in \pi_{2t+k+1}(S^{2t+1})$ the e-invariant of the r-th Hopf invariant satisfies the formula

(1.7) $\quad e_{\mathbb{C}}\gamma_r(\alpha) = (-1)^{r+1} r^{t-1}(r^{(k+1)/2} - 1)e_{\mathbb{C}}(\alpha)$.

Theorem (1.3) shows that this formula is only relevant if $r = p^\nu$ is a prime power. Using results of B. Gray [19] we derive from (1.7),

(1.8) **Proposition.** Let $\alpha \in \pi_{2m}(S^{2t+1})$. Then $e_{\mathbb{C}}\gamma_r(\alpha) \neq 0$, $r > 1$, implies $r = 2^\nu$ or $r = p$ is an odd prime.

(1.9) **Theorem.** Let p be an odd prime and assume $k > t \geq 1$ are given with $p^{t-1} | k$. Then there exists an element

$$\alpha \in \pi_{2k(p-1)+2t}(S^{2t+1})$$

of order p^t with e-invariant

IV

$$e_{\mathbb{C}}(\alpha) = -\frac{1}{p^t} \mod \mathbb{Z}$$

such that the p-th Hopf invariant

$$\gamma_p(\alpha) \in \pi_{2k(p-1)+2t}(S^{2pt+1})$$

is an element of order p with e-invariant

$$e_{\mathbb{C}}\gamma_p(\alpha) = -\frac{1}{p} \mod \mathbb{Z} .$$

Proof. The element α is constructed by B. Gray in [19]. He proves that $e_{\mathbb{C}}(\alpha) = -1/p^t \mod \mathbb{Z}$. Using (1. 7) we see that $e_{\mathbb{C}}\gamma_p(\alpha) = -1/p \mod \mathbb{Z}$. By a result of F. R. Cohen, J. C. Moore and J. A. Neisendorfer the p-component of $\pi_*(S^{2t+1})$ has exponent p^t. Therefore α is in fact an element of order p^t. From Theorem (1. 3) we know that $\gamma_p(\alpha)$ is an element of order p. //

In other words, (1. 9) shows that elements α, which are in the image of the J-homomorphism, have non trivial higher order Hopf invariants on their sphere of origin.

Remark. The examples above yield further examples by considering $[j_t, j_t] \circ \alpha$ where $[j_t, j_t]$ is the Whitehead square of a generator $j_t \in \pi_{t+1}(S^{t+1})$, t odd. In III (6. 6) we prove for $\alpha \in [\Sigma S^n, S^{2t+1}]$ the formula

$$\gamma_{2p}([j_t, j_t] \circ \alpha) = (2^p j_{2pt}) \circ \gamma_p(\alpha) .$$

§ 2. Proof of theorem (1. 3)

We first consider the James-Hopf invariants γ_n which are defined with respect to the lexicographical ordering from the left.

Since X is a Co-H-space we know that all geometric cup products are trivial. This fact and formula III (4. 2) yield equations for higher order Hopf invariants, from which we will derive the proposition.

We know that for binomial coefficients the greatest common divisor

$$\Lambda_m = \gcd\{\binom{m}{i} \mid 0 < i < m\}$$

is given by

(1) $$\Lambda_m = \begin{cases} p & \text{if } m = p^\nu,\ p \text{ a prime, } \nu \geq 1 \\ 1 & \text{otherwise.} \end{cases}$$

For the numbers

(2) $$N = N^k_{n,m}, \quad P = P^k_{n,m}$$

in III (4. 2) we get from III (4. 2) and III (4. 5)

(3) $$(Nj)\gamma_{n+m}(\alpha) = -P[j,\ j]\gamma_{2(n+m)}(\alpha)$$

Here we know by III (4. 5) that the term $[[j,\ j],\ j]\gamma_{3(n+m)}(\alpha)$ in III (4. 2) is trivial if $(n + m)k$ is odd.

Using III (1. 9) we deduce from (3)

(4) $$N \cdot \gamma_{n+m}(\alpha) = (L[j,\ j])\gamma_{2(n+m)}(\alpha)$$

with

$$L = -P - \frac{N(N-1)}{2}$$

Let $R = (n + m)$ and $r + s = 2R$. Then (4) implies

(5) $$N^k_{r,s} N^k_{n,m} \gamma_R(\alpha) = 0.$$

If k is even we thus get from III (4. 3) and (1)

(6) $$\Lambda_{2R} \cdot \Lambda_R \gamma_R(\alpha) = 0.$$

Thus $4\gamma_R(\alpha) = 0$ if $R = 2^\nu$ and $p\gamma_R(\alpha) = 0$ if $R = p^\nu$, p an odd prime, and $\gamma_R(\alpha) = 0$ otherwise.

Now let $R = n + m = 2$. Then we know by III (6. 2) $[j,\ j]\gamma_4(\alpha) = 0$ and therefore by (4) with $n = m = 1$.

(7) $$2\gamma_2(\alpha) = 0 \text{ if } |\alpha| = k \text{ even}$$

Moreover we have by III (5. 2)

(8) $\gamma_2\gamma_{2^n}(\alpha) = N^k_{2^n-1,\,2^n}\gamma_{2^{n+1}}(\alpha)$

where $|\alpha| = k$ can be odd or even. Now

$$N^k_{2^n-1,\,2^n} = \begin{cases} \binom{2^{n+1}-1}{2^n-1} & k \text{ even} \\ \binom{2^n-1}{2^{n-1}-1} & k \text{ odd} \end{cases}$$

is always an odd number. Therefore by (7)

(9) $2\gamma_{2^n}(\alpha) = 0$ for $n > 1$, $|\alpha|$ odd or even.

Thus the proposition is proved for k even. Now let k be odd. Then (5) and III (4. 3) yield for $R \geq 3$

(10) $\Lambda_R\,\Lambda_{[R/2]}\,\gamma_R(\alpha) = 0.$

If R is even, $R \geq 4$, this implies $p\gamma_R(\alpha) = 0$ if $R = 2p^\nu$ ($\nu \geq 1$ and p an odd prime) and $4\gamma_R(\alpha)$ if $R = 2^\nu$, ($\nu \geq 2$) and $\gamma_R(\alpha) = 0$ for other even R. This and (9) proves the proposition for $\gamma_R(\alpha)$ with odd k and even R. We still have to show that $\gamma_R(\alpha) = 0$ if R and k are odd.

Assume now R and k are odd. From III (4. 5) we see

$$\gamma_R(\alpha) = \binom{R-2}{(R-1)/2}\,[j,\ j]\gamma_{2R}(\alpha)\,.$$

Using (10) for $2R$ instead of R we have

(11) $\Lambda_{2R}\Lambda_R\gamma_R(\alpha) = 0$ for Rk odd.

From this equation we derive by (1) $p\gamma_R(\alpha) = 0$ if $R = p^\nu$, p an odd prime and $\gamma_R(\alpha) = 0$ for other odd R. In case $p\gamma_R(\alpha) = 0$, $R = p^\nu$ odd, we can assume α has odd order. We therefore know

(12) $\alpha = \Sigma\alpha' + [j,\ j] \circ \alpha''$

since k is odd. From III (5. 3) and III (6. 6) we get

(13) $\gamma_R(\alpha) = \gamma_R([j, j]\alpha'') = 0.$

Now the proposition is proved for James Hopf invariants with respect to the lexicographical ordering.

Proposition. Let X be a finite dimensional Co-H-space. Then for $n \geq 2$ the set Λ_n^k in III (2. 4) contains exactly one function $\gamma_n : [\Sigma X, \Sigma S^k] \to [\Sigma X, \Sigma S^{nk}]$, that is, all possible definitions of James-Hopf invariants coincide for $n \geq 2$.

Proof. If nk is even we have for $\lambda \in \Lambda_n^k$

(14) $\lambda(\alpha) = \gamma_n(\alpha) + s[j, j]\gamma_{2n}(\alpha).$

$\gamma_{2n}(\alpha)$ might be non trivial for $n = 2^m$. Here we consider only $m \geq 1$ since $n \geq 2$. From III (6. 2) and (8) we can derive

(15) $[j, j]\gamma_{2^m}(\alpha) = 0,$ $m \geq 2$ and k even or odd.

Now let nk be odd. Then we have for $\lambda \in \Lambda_n^k$ the formula

(16) $\lambda(\alpha) = s[j, j]\gamma_{2n}(\alpha) + t[[j, j], j]\gamma_{3n}(\alpha)$

Here $[[j, j], j]\gamma_{3n}(\alpha)$ is trivial by III (4. 5) and III (1. 1). Moreover $\gamma_{2n}(\alpha)$ might be non trivial for $n = p^\nu$, p odd. However, for $R = p^\nu$, p and k odd we know

(17) $0 = \gamma_R(\alpha) = \binom{R-2}{(R-1)/2} [j, j]\gamma_{2R}(\alpha)$

by III (4. 5). Since for $R = p^\nu$, p an odd prime,

$$\binom{R-2}{(R-1)/2} \not\equiv 0 \mod p$$

we can derive from (17) $[j, j]\, \gamma_{2R}(\alpha) = 0$. This proves $\lambda(\alpha) = 0$ for nk odd. //

§ 3. Zassenhaus terms for an odd prime p

As an application of the results in chapters II and III we give in this section proofs of the purely algebraic results on Zassenhaus terms in I, § 4.

Proof of I (4. 8) for $\nu = 1$. Let $w = x^1 \dots x^k \in FM(\{x, -x \mid x \in X\})$. For $\underline{\alpha}_p$ in (1. 6) we consider

$$\sum_{i=1}^{k} x^i \circ \underline{\alpha}_p \in [\Sigma P^{2p}(p), \bigvee_{x \in X} \Sigma S^2] = G$$

where $x \in X$ denotes as well the inclusion of ΣS^2 for the index x. We know by the general left distributivity law II (2. 8)

$$\sum_{i=1}^{k} x^i \circ \underline{\alpha}_p = (\sum_{i=1}^{k} x^i) \circ \underline{\alpha}_p + c_p(w) \circ \gamma_p(\underline{\alpha}_p) .$$

Since $(-1) \circ \underline{\alpha}_p = \underline{\alpha}_p \circ (-1)$ by III (1. 9) and since G is an abelian group we have for $v = y^1 \dots y^s \in FM(\{*, -x \mid x \in X)$ with $\pi v = \pi w$ also

$$\sum_{i=1}^{k} x^i \circ \underline{\alpha}_p = \sum_{i=1}^{s} y^i \circ \underline{\alpha}_p .$$

Therefore

$$c_p(w) \circ \gamma_p(\underline{\alpha}_p) = c_p(v) \circ \gamma_p(\underline{\alpha}_p) .$$

Since $\gamma_p(\underline{\alpha}_p) : \Sigma P^{2p}(p) \to \Sigma S^{2p}$ is essentially the pinch map it follows

$$c_p(w) \equiv c_p(v) \mod p. \quad //$$

Proof of I (4. 9). Let $J_i = J_i(S^2)$ be the i-fold reduced product of the 2-sphere. We consider the group

$$G = [\Sigma(S^2 \times J_i), \ \Sigma(S^2 \vee S^2 \vee S^2)] .$$

Let x, y, z be the three inclusions of ΣS^2 into $\Sigma(S^2 \vee S^2 \vee S^2)$.

Let p, q be the projections of $\Sigma(S^2 \times J_i)$ onto ΣJ_i and ΣS^2 respectively. Moreover let $\bar{g} : \Sigma J_i \to \Sigma S^2$ be the retraction given by the adjoint of g in II (2. 1). Instead of I (4. 11) we now consider the commutator

$$B = [z \circ e,\ x \circ \eta + y \circ \eta]$$

in the group G, where $\eta = \bar{g} \circ p$ and where $e = q$. We know that we have a homotopy equivalence

$$\Sigma(S^2 \times J_i) \overset{R}{\simeq} \bigvee_{j=1}^{i} S^{2j} \vee \bigvee_{j=1}^{i} S^{2j+2} = W$$

Let I_j be the inclusion of S^{2j} and J_j the inclusion of S^{2j+2} into W. Then we set

$$R = \sum_{j=1}^{i} I_j \circ \gamma_j(\eta) + \sum_{j=1}^{i} J_j \circ (e \cup \gamma_j(\eta))$$

We now fix an odd prime p and we set $i = p^\nu$, $\nu \geq 1$. Then R yields the projection

$$r : G \to [\Sigma S^{2p^\nu+2},\ \Sigma(S^2 \vee S^2 \vee S^2)] \otimes \mathbb{Z}/p\mathbb{Z}$$

with $r(x) = (J^*_{p^\nu})(R^*)^{-1}(x) \otimes 1$. We calculate $r(B)$ in two different ways: First we know by I (1.14) that

$$B = [z \circ e,\ x \circ \eta] + [z \circ e,\ y \circ \eta] + [[z \circ e,\ x \circ \eta],\ y \circ \eta]$$

We expand $[z \circ e,\ x \circ \eta]$ and $[z \circ e,\ y \circ \eta]$ by II (3.4), where we use the equation in the proof of II (1.12). We obtain

$$[z \circ e,\ x \circ \eta] = \sum_{n \geq 1} [[z,\ x^n] \circ (e \cup \gamma_n(\eta)) = U.$$

We moreover can expand by I (1.15) and II (3.4) the term $[U,\ y \circ \eta]$, so that we get

$$[U,\ y \circ \eta] = \sum_{n \geq 1} U_n \circ (e \cup \gamma_n(\eta))$$

for certain U_n. U_n is obtained by the rules for the cup product and for Hopf invariants γ_n in III. Since $\binom{p^\nu}{i}$ for $i = 1, \ldots, p^\nu - 1$ is divisible by p we see that U_{p^ν} is divisible by p. Moreover, since the Hopf invariants of $e \cup \gamma_n(\eta)$ vanish, we see by II (2.8)

IV

$$B = \sum_{n\geq 1} ([[z, x^n] + [[z, y^n] + U_n) \circ (e \cup \gamma_n(\eta)) .$$

Therefore we obtain

$$r(B) = [[z, x^{p^\nu}] + [[z, y^{p^\nu}] .$$

On the other hand we have by II (2. 8)

$$B = [z \circ e, (x + y) \circ \eta + \sum_{n\geq 2} c_n(x, y) \circ \gamma_n(\eta)]$$

If we expand this term by I (1. 15) and by II (3. 4) we see by considerations as above

$$r(B) = [[z, (x + y)^{p^\nu}] + \sum_{\substack{u+w=v \\ u\geq 0 \\ w\geq 1}} [[z, c_{p^w}(x, y)^{p^u}] .$$

This proves equation I (4. 14), since we have the embedding

$$L_{\mathbb{Z}}(x, y, z) \otimes \mathbb{Z}/p\mathbb{Z} \hookrightarrow \pi_*(\Omega\Sigma(S^2 \vee S^2 \vee S^2)) \otimes \mathbb{Z}/p\mathbb{Z}$$

With the argument I (4. 15) the proof of I (4. 9) is complete. //

PART B: HOMOTOPY THEORY OVER A SUBRING R OF THE RATIONALS $\mathbb{Q}$ WITH $1/2, 1/3 \in R$

V. THE HOMOTOPY LIE ALGEBRA AND THE SPHERICAL COHOMOTOPY ALGEBRA

In chapter III we gave a list of equations for homotopy operations on homotopy groups $[\Sigma X, \Sigma S^n]$ of spheres. We simplify these equations if we localize with respect to a subring R of the rationals $\mathbb{Q}$ containing $1/2, 1/3 \in R$. In these localized groups $[\Sigma X, \Sigma S^n]_R$ the terms which are compositions with the Whitehead products $[j, j]$ (n even) and with $[[j, j], j]$ vanish, $j = j_n \in \pi_{n+1}(\Sigma S^n)$.

This way we are led to introduce the spherical cohomotopy functor, which associates with a connected space X an algebra with divided powers $M(X, R)$. This functor has properties dual to the properties of the homotopy functor, which associates to a connected loop space ΩY the Lie algebra $L(Y, R) = \pi_*(\Omega Y) \otimes R$.

§ 0. Notation

We recall some notations:

A group G is nilpotent if there exists an integer $k \geq 1$ such that an iterated commutator of any k of its elements taken in any order is zero. G. W. Whitehead proved that $[X, \Omega Y]$ is a nilpotent group if X is finite dimensional. For such a nilpotent group we have the Malcev completion or rationalization $[X, \Omega Y]_{\mathbb{Q}}$ and more generally the localization $[X, \Omega Y]_R$ with respect to any subring $R \subset \mathbb{Q}$. We assume X and ΩY to be connected and $1/2, 1/3 \in R$.

(0.1) Definition. Let $R \subset \mathbb{Q}$ be a subring of the rationals. We say a nilpotent group G is an R-local group if G is uniquely divisible with respect to R, that is the function $x \mapsto x^n$, $x \in G$, is bijective for all $n \neq 0$ with $1/n \in R$. A $\mathbb{Q}$-local group is also called a rational group. For each nilpotent group G the R-localization $G \to G_R$ is given where G_R is an R-local group. It has the universal property: Any group homomorphism $G \to H$ into an R-local group factors in an unique way over

V

$G \to G_R$. $G_{\mathbb{Q}}$ is called the rationalization or Malcev completion of G. Compare [25]. Similarly, for a nilpotent space X the R-localization $X \to X_R$ is defined [25].

(We remark that R-localization here is π-localization in [25] with π the set of primes <u>not</u> invertible in R.)

The R-localization $X \to X_R$ induces isomorphisms of R-local groups

$$(0.2) \quad [\Sigma X, Y]_R \cong [\Sigma X, Y_R] \cong [\Sigma X_R, Y_R].$$

Since we assume ΩY and X to be connected, the localizations Y_R and $\Sigma X_R = (\Sigma X)_R$ are defined. A particular case of (0.2) is

$$(0.3) \quad \pi_n(\Omega Y) \otimes R = [\Sigma S^n, Y_R] = [\Sigma S^n_R, Y_R]$$

for $n \geq 1$.

If x is an element of a graded R-module, then $|x|$ denotes its degree.

(0.4) Definition. Compare [15]. A <u>Lie algebra</u> L is a (positively graded) R-module with R bilinear pairings

$$[\ ,\ ] : L_n \times L_m \to L_{n+m}$$

which satisfy the relations of

(i) antisymmetry $[x, y] = -(-1)^{|x||y|}[y, x]$ for all x and y in L

(ii) the Jacobi identity:

$$[x, [y, z]] = [[x, y], z] + (-1)^{|x||y|}[y, [x, z]]$$

for all x, y, and z in L.

We consider only connected Lie algebras, i.e. those with $L_0 = 0$. Let Lie_R be the <u>category of connected Lie algebras.</u>

(0.5) Definition. Compare [41]. We say a graded module $A = \{A^n, n \geq 0\}$ over R is a (connected graded commutative) <u>algebra</u> over R if $A^0 = R$ and if an associative multiplication

$$\cup : A \otimes_R A \to A$$

with unit $1 \in R = A^0$ is given which is commutative, that is $x \cup y = (-1)^{|x||y|} y \cup x$. We say such an algebra has divided powers γ if functions

$$\gamma_r : A^n \to A^{nr}, \qquad n \geq 1, \quad r \geq 0,$$

are given satisfying the following set of axioms: $x, y \in A$

(a) $\gamma_0 x = 1$ and $\gamma_1 x = x$

(b) $\gamma_r x = 0$ for $r > 1$ and $|x|$ odd

(c) $\gamma_r(x+y) = \sum_{\substack{i+j=r \\ i,j\geq 0}} \gamma_i(x) \cup \gamma_j(y)$

(d) $\gamma_n(x \cup y) = x^n \cup (\gamma_n y) = (\gamma_n x) \cup y^n$

(e) $\gamma_n \gamma_m x = \dfrac{(m \cdot n)!}{(m!)^n (n!)} \gamma_{m \cdot n}(x)$

(f) $\gamma_n(x) \cup \gamma_m(x) = \binom{n+m}{n} \gamma_{n+m}(x)$

From (d) follows

(g) $\gamma_n(\lambda y) = \lambda^n \gamma_n(y)$ for $\lambda \in R$

and from (f) we obtain

(h) $n!\, \gamma_n(x) = x^n$.

$m! = 1\, 2 \ldots m$, and $\binom{m}{n} = m!/(n!(m-n)!)$ is the binomial coefficient.

Let $\mathrm{Div\ alg}_R$ be the category of algebras with divided powers. Morphisms are the algebra homomorphisms f of degree 0 compatible with γ, that is $f \circ \gamma_r = \gamma_r \circ f$.

(0.6) **Remark.** If A is uniquely divisible equation (h) determines γ, namely $\gamma_n(x) = \frac{1}{n!} x^n$. Clearly the function $\frac{x^n}{n!}$ satisfies all the axioms. So we see, the rational cohomology $H^*(X, \mathbb{Q})$ of a connected space X is an algebra with divided powers over $\mathbb{Q}$.

§ 1. The homotopy Lie algebra and the spherical cohomotopy algebra

Let Top_0 and Top_1 be the homotopy category of connected and 1-connected CW-spaces respectively, (clearly with base points). The homotopy Lie

V
<u>algebra</u> is the functor

(1.1) $L(\,.\,, R) : Top_1 \to Lie_R$

which associates with a simply connected space Y the Lie algebra

$$L(Y, R) = \pi_*(\Omega Y) \otimes R .$$

The R-local spheres S^n_R, $n \geq 1$, are the universal objects for this functor since we have

$$\pi_n(\Omega Y) \otimes R = [S^n_R, \Omega Y_R] .$$

The Lie bracket is the Samelson product. Equation II (1.7) and II (1.8) show that the Samelson product satisfies the equations (0.4) of a Lie algebra.

Rational homotopy theory of Quillen [34] and Sullivan [40] shows that the rational cohomology functor $H^*(\,.\,, \mathbb{Q})$ has properties which are dual to the properties of the functor $L(\,.\,, \mathbb{Q})$.

More generally we now exhibit for a subring $R \subset \mathbb{Q}$ $(1/2, 1/3 \in R)$ the functor $M(\,.\,, R)$ which is dual to $L(\,.\,, R)$. This extends the above duality in rational homotopy theory, we have $M(\,.\,, \mathbb{Q}) = H^*(\,.\,, \mathbb{Q})$. The functor

(1.2) $M(\,.\,, R) : Top_0 \to Div\ alg_R$

associates with a connected space X an algebra with divided powers $M(X, R)$, see (0.5), which we call the spherical <u>cohomotopy algebra</u> of Y. The universal objects for $M(\,.\,, R)$ are the spaces

(1.3) $$\Omega^n_R = \begin{cases} \Omega\Sigma S^n_R & n \text{ even} > 0 \\ S^n_R & n \text{ odd} . \end{cases}$$

For $R = \mathbb{Q}$, it is well known that

(1.4) $\Omega^n_{\mathbb{Q}} = K(\mathbb{Q}, n)$

is the Eilenberg-MacLane space of $\mathbb{Q}$ in dimension n.

If X is finite dimensional (see II (1.11)) $M^n(X, R)$ is defined as a set by $M^0(X, R) = R$ and for $n > 0$ by

$$(1.5) \quad M^n(X, R) = [X, \Omega^n_R] = \begin{cases} [\Sigma X, \Sigma S^n_R], & n \text{ even} > 0 \\ [X, S^n_R], & n \text{ odd}. \end{cases}$$

If X is not finite dimensional

$$M^n(X, R) = \varprojlim M^n(X^N, R)$$

is the inverse limit given by the inclusions $X^0 \subset X^1 \subset \ldots$ of finite dimensional skeletons of a CW-model of X, see (1.4) in [11]. This graded set $M(X, R) = \{M^n(X, R)\}_{n \geq 0}$ has in a natural way the algebraic structure of an algebra with divided powers which we define as follows.

For $x \in M^n(X, R)$ we set

$$(1.6) \quad \tilde{x} = \begin{cases} x & \text{if } n \text{ is even} \\ \Sigma x & \text{if } n \text{ is odd} \end{cases}$$

so that $\tilde{x} \in [\Sigma X, \Sigma S^n_R]$ for all n. As we know, the suspension map (n odd)

$$\Sigma : M^n(X, R) = [X, S^n_R] \to [\Sigma X, \Sigma S^n_R]$$

is injective, since $\frac{1}{2} \in R$. Therefore x is uniquely determined by $\tilde{x}$.

Let j_n be the identity of ΣS^n_R. For each $\lambda \in R$ we have the map $\lambda j_n : \Sigma S^n_R \to \Sigma S^n_R$ of degree λ. Furthermore we know $S^n_R \wedge S^m_R = S^{n+m}_R$. With these notations we define for $x, y \in M^*(X, R)$, $\lambda \in R$:

(i) $(\lambda x)^\sim = (\lambda j_n) \circ \tilde{x}$

(ii) $(x + y)^\sim = \tilde{y} + \tilde{x} + [j, j] \left(\frac{\tilde{x} \cup \tilde{y}}{2}\right)$ where $j = j_n$ for $|x| = |y| = n$. Clearly $[j, j] = 0$ if n is even.

(iii) $(x \cup y)^\sim = \tilde{x} \underline{\cup} \tilde{y}$.

(1.7) Theorem. For finite dimensional X, (i) and (ii) determine an R-module structure on $M^*(X, R)$ and (iii) yields an algebra multiplication $\cup$ on $M^*(X, R)$. Moreover the James-Hopf invariants

$$\gamma_r : M^n(X, R) \to M^{rn}(X, R)$$

(n even) are divided powers on this algebra. If X is not finite dimensional $M^*(X, R)$ is the inverse limit of the algebras with divided powers $M^*(X^N, R)$, see (1.5).

(1.8) **Remark.** There are various possibilities to define the James Hopf invariants. However, since $\frac{1}{2} \in R$ and since n is even all possible orderings for the definition of γ_r yield the same function, see III § 2.

(1.9) **Remark.** P. S. Selick has shown that S^n_R for n odd and $1/2, 1/3 \in R$ is an H-space which is homotopy associative and homotopy abelian [15]. In fact such an H-multiplication μ is given by defining

$$\mu = p_1 + p_2 \in M^n(S^n_R \times S^n_R, R)$$

where p_1 and p_2 are the projections $S^n_R \times S^n_R \to S^n_R$. Clearly, μ induces the abelian group structure on $M^n(X, R)$.

Proof of (1.7). It is enough to prove (1.7) for finite dimensional X. We first show that $M^n(X, R)$ is an R-module. If n is even, ΣS^n_R is an H-space and the multiplication on ΣS^n_R induces the abelian group structure on $M^n(X, R)$. Thus $M^n(X, R)$ is indeed an R-module in this case. The case, n odd, is more complicated. First we have to show that $(x + y)^\sim$ in (ii) is desuspendable. For this it suffices to prove that the James-Hopf invariant $\gamma_2(x + y)^\sim$ vanishes, since it is a result of James, see Ann. Math. 65 (1957) 74-106, that

$$S^n_R \hookrightarrow J(S^n_R) \xrightarrow{g_2} J(S^{2n}_R)$$

is a fiber sequence which admits a homotopy section if $\frac{1}{2} \in R$. Using III (5.3) and III (6.6) we see

$$\gamma_2(x + y)^\sim = \tilde{y} \cup \tilde{x} + \tilde{x} \cup \tilde{y} = 0$$

where we use the familiar fact $\gamma_2([j, j]) = 2$. Moreover (i) and (ii) are compatible for n odd since by III (4.6) we have $\tilde{x} \cup \tilde{x} = 0$.

It remains to prove that (ii) defines an abelian group structure.

In fact by III (1. 3) and III (3. 1) (since $1/3 \in R$):

$$\tilde{y} + \tilde{x} + [j, j]\left(\frac{\tilde{x} \cup \tilde{y}}{2}\right) = \tilde{x} + \tilde{y} + [j, j]\left(\frac{\tilde{y} \cup \tilde{x}}{2}\right)$$

Furthermore we show that (ii) is associative.

$$((x + y) + z)^{\sim} = \tilde{z} + (x + y)^{\sim} + [j, j](\tfrac{1}{2}(x \cup y)^{\sim} \cup \tilde{z})$$

$$= \tilde{z} + \tilde{y} + \tilde{x} + [j, j]\tfrac{1}{2}(\tilde{x} \cup \tilde{y} + \tilde{x} \cup \tilde{z} + \tilde{y} \cup \tilde{z}).$$

On the other hand

$$(x + (y + z))^{\sim} = (y + z)^{\sim} + \tilde{x} + [j, j](\tfrac{1}{2}\tilde{x} \cup (y + z)^{\sim})$$

$$= \tilde{z} + \tilde{y} + [j, j]\tfrac{1}{2}(\tilde{y} \cup \tilde{z}) + \tilde{x} + [j, j]\tfrac{1}{2}(\tilde{x} \cup \tilde{y} + \tilde{x} \cup \tilde{z}).$$

Thus using III (1. 3) we have associativity, since $1/3 \in R$.

That $\cup$ is an algebra multiplication follows from (1. 15), (1. 17) and (1. 22) in I and from (3. 1) in III. Property (c) and (d) in (1. 7) are proved in (2. 14) and (2. 15) in II. (e) and (f) follow from (4. 2) and (5. 2) in III. //

The following corollary corresponds to IV (1. 3)

(1. 10) Corollary. If X is a Co-H-space the Hopf invariant $\gamma_r : M^n(X, R) \to M^{rm}(X, R)$ is a homomorphism of R-modules, also $p \cdot \gamma_r(x) = 0$ if $r = p^\nu$ is a prime power and $\gamma_r(x) = 0$ otherwise.

As we saw in IV for a prime p and Co-H-space X the Hopf invariant γ_p on $M^m(X, R)$ need not be trivial. For example for a sphere S^N

(1. 11) $\gamma_p : M^{2t}(S^N, R) \to M^{2tp}(S^N, R)$

is non trivial if $N = 2k(p - 1) + 2t - 1$ where p^{t-1} divides k, $k > t \geq 1$, $1/p \notin R$.

Proof of (1. 10). Since X is a Co-H-space the reduced diagonal $\tilde{\Delta} : X \to X \wedge X$ is nullhomotopic. Therefore all cup products in $M^*(X, R)$ are trivial. Now if A is an algebra with divided powers and if the cup product in A is trivial the operation γ_r is a homomorphism of groups

V

by (c) in (1. 7). Moreover,

$$p\gamma_r(x) = 0 \text{ if } r = p^v, \; v \geq 1, \; p \text{ a prime}$$

$$\gamma_r(x) = 0 \text{ for all other } r.$$

This follows from (f) in (1. 7) since the greatest common divisor of $\binom{r}{n}$ for $0 < n < r$ is equal to p if $r = p^v$ and is 1 otherwise. //

§ 2. Homotopy groups of spheres and homotopy coefficients

It is well known that for the p-primary component (p odd) of homotopy groups of even dimensional spheres we have a splitting

$$\pi_{n+1}(S^{2t})_p = \pi_n(S^{2t-1})_p \oplus \pi_{n+1}(S^{4t-1})_p.$$

Thus only homotopy groups of odd dimensional spheres are relevant. We here describe the algebraic structure of primary homotopy operations on these groups.

The graded R-modules $M^*(X, R)$ and $\pi_*(\Omega Y) \otimes R$ have additional structure in that homotopy groups of spheres operate on them. We call the double graded R-module $M = M_R^{**}$ with $m, n \geq 1$

(2. 1) $$M_R^{m,n} = M^m(S^n, R) = \begin{cases} [\Sigma S_R^n, \Sigma S_R^m] & m \text{ even} \\ [S_R^n, S_R^m] & m \text{ odd} \end{cases}$$

the coefficients of the functors $M^*(. , R)$ and $\pi_*(\Omega \, . \,) \otimes R$. We have an isomorphism

(2. 2) $$\pi_{n+1}(\Sigma S_R^m) \underset{\psi}{\cong} \begin{cases} M_R^{m,n} & m \text{ even} \\ M_R^{m,n} \oplus M_R^{2m,n} & m \text{ odd} \end{cases}$$

where $\psi(\alpha, \beta) = \Sigma\alpha + [j, j] \circ \beta$ if m is odd.

The module of coefficients $M = M^{**}$ in (2. 1) has the following algebraic structure:

We know by (2. 2)

(i) $M^{**} = \{M^{m,n}, \; m, n \geq 1\}$ is a double graded R-module with

$$M^{m,n} = \begin{cases} 0 & m > n \\ R & m = n \\ \text{finite group} & m < n . \end{cases}$$

(ii) We have a bilinear pairing $(m, n, k \geq 1)$

$$\odot : M^{m,n} \otimes_R M^{n,k} \to M^{m,k}$$

defined by $(\zeta \odot \eta)^\sim = \tilde{\zeta} \circ \tilde{\eta}$, see (1.6)

(iii) We have a bilinear pairing $(m, n, m', n' \geq 1)$

$$\# : M^{m,n} \otimes_R M^{m',n'} \to M^{m+m',n+n'}$$

defined by $(\zeta \# \eta)^\sim = \tilde{\zeta} \underline{\#} \tilde{\eta} = \tilde{\zeta} \# \tilde{\eta}$, see II (1.14).

(iv) We have homomorphisms of R-modules $(m, r, n \geq 1)$

$$\gamma_r : M^{m,n} \to M^{mr,n}$$

which are the Hopf invariants.

(v) We have elements $(n \geq 1)$

$$e^n = 1 \in R = M^{n,n}$$

defined by $(e^n)^\sim = j_n$.

These operations $(\odot, \#, \gamma_r, e^r)$ on M^{**} satisfy the following relations (a, b, ..., i)

(2.3) **Definition.** Let R be a subring of $\mathbb{Q}$. We call any double graded R-module $M = M^{**}$ with operations $(\odot, \#, \gamma_r, e^r)$ as in (i)... (v) above a module of homotopy coefficients if

(a) $\odot$ is associative and has units e^n,

$$e^m \odot \zeta = \zeta = \zeta \odot e^n \quad \text{for} \quad \zeta \in M^{m,n}$$

(b) $\#$ is associative and is commutative in the sense

$$\zeta \# \eta = (-1)^{mm'+nn'} \eta \# \zeta$$

(c) $\overset{\vee}{e}^n \# e^m = e^{n+m}$

(d) $\zeta \# \eta = (e^m \# \eta) \odot (\zeta \# e^{n'})$

$= (\zeta \# e^{m'}) \odot (e^n \# \eta).$

(e) For e^1 the homomorphism of R-modules

$$e : M^{m,n} \to M^{m+1,n+1}, \quad e(\zeta) = e^1 \# \zeta,$$

is an isomorphism for $n < 2m$. Moreover e maps an element of prime power order p^v to an element of order p^v or p^{v-1}.

(f) $e(\zeta \odot \eta) = (e\zeta) \odot (e\eta)$.

(g) For $\gamma_r : M^{m,n} \to M^{mr,n}$ we have $p \cdot \gamma_r(\zeta) = 0$ if $r = p^v$ is a prime power and m is even. Otherwise $\gamma_r(\zeta) = 0$. Moreover $\gamma_r(\zeta) = 0$ for $\zeta \in$ Image e.

(h) $$\gamma_n \gamma_m \zeta = \frac{(m \cdot n)!}{(m!)^n n!} \gamma_{m \cdot n}(\zeta)$$

(i) $$\gamma_n(\zeta \odot \eta) = \sum_{r \geq 1} \Gamma_n^r(\zeta) \odot \gamma_r(\eta) \quad \text{with}$$

$$\Gamma_n^r(\zeta) = \sum_{\substack{i_1 + \dots + i_r = n \\ i_1, \dots, i_r \geq 1}} \gamma_{i_1}(\zeta) \# \dots \# \gamma_{i_r}(\zeta)$$

(2.4) **Remark.** For M^{**} in (2.1) the homomorphism e in (e) is the double suspension on $M^{m,n}$, m odd, and is the identity if n is even. The double suspension was studied by Toda and in [15]. There it is shown that e has the properties described in (e) above. That e is an isomorphism for $n < 2m$ is the Freudenthal suspension theorem if m is odd. All other properties of M^{**} as listed in (a) ... (i) are proved in chapter III. We call M^{**} in (2.1) the module of spherical coefficients. This module is only defined for $1/2, 1/3 \in R$.

(2.5) **Remark.** By use of (d) and (b) we see that the homomorphism e and the pairing $\odot$ determine #. Moreover from (g) and (h) we derive that the invariants γ_p (p a prime) determine all other invariants γ_r. Moreover (d) and (i) show that γ_r vanishes on products $\zeta \# \eta$. Clearly we have $e^n \# \zeta = e \dots e(\zeta)$.

(2. 6) **Remark.** Modules of homotopy coefficients form a category. The morphism $M \to M'$ are of bidegree (0, 0) and are compatible with all operations. For example $\underline{R}$ with

$$\underline{R}^{m,n} = \begin{cases} 0 & \text{for } m \neq n \\ R & \text{for } m = n \end{cases}$$

is in a canonical way a module of homotopy coefficients and we have for each M canonical morphisms

$$\underline{R} \overset{i}{\hookrightarrow} M \overset{\varepsilon}{\to} \underline{R}$$

of modules of homotopy coefficients. The module of spherical coefficients is the attractive object in this category of homotopy coefficients. It is the main problem of homotopy theory to find additional algebraic properties which characterize this object uniquely.

(2. 7) For $R = \mathbb{Q}$ we have an isomorphism $M_{\mathbb{Q}}^{**} \cong \underline{\mathbb{Q}}$ of coefficient modules.

We now describe the operation of M_R^{**} on the functor $M^*(\,.\,, R)$. More generally than (ii) in (2. 3) there is the bilinear pairing of R-modules.

(2. 8) $$M_R^{m,n} \otimes_R M^n(X, R) \overset{\odot}{\to} M^m(X, R)$$

defined by $(\zeta \odot x)^\sim = \tilde{\zeta} \circ \tilde{x}$, see (1. 6).

This pairing has the following algebraic structure:

(2. 9) **Definition.** Let A be an algebra with divided powers, see (0. 5). We say A is an M-algebra if for homotopy coefficients $M = M_R^{**}$ we have a bilinear pairing of R-modules $(m, n \geq 1)$

$$M_R^{m,n} \otimes_R A^n \overset{\odot}{\to} A^m$$

with the following properties $(\zeta, \eta \in M, x, y \in A)$

(a) $e^n \odot x = x$ for the elements $e^n \in M^{n,n} = R$

(b) $(\zeta \odot \eta) \odot x = \zeta \odot (\eta \odot x)$

(c) $(\zeta \odot x) \cup (\eta \odot y) = (\zeta \# \eta) \odot (x \cup y)$.

(d) $\gamma_n(\zeta \odot x) = \sum_{r \geq 1} \Gamma_n^r(\zeta) \odot \gamma_r(x)$, see (2. 3) (i), $n \geq 1$.

Let Div alg_M be the category of M-algebras. Morphisms are the homomorphisms $f : A \to A'$ in Div alg_R with $f(\zeta \odot x) = \zeta \odot f(x)$ for $\zeta \in M$, $x \in A$, see (0. 5).

(2. 10) Proposition. With the structure (2. 8) the spherical cohomotopy algebra (1. 2) is a contravariant functor

$$M^*(\ldots, R) : \text{Top}_0 \to \text{Div alg}_M.$$

Proof. On maps $f : X' \to X$ we define

$$f^*(x)^\sim = \tilde{x} \circ (\Sigma f)$$

for $x \in M^*(X, R)$, compare VI (4. 1). //

The coefficients M_R^{**} operate also on the homotopy functor $\pi_*(\Omega\ .\ ) \otimes R$, that is we have a pairing

(2. 11) $$(\pi_m(\Omega Y) \otimes R) \times M_R^{m,n} \xrightarrow{\odot} \pi_n(\Omega Y) \otimes R$$

defined by $(\alpha \odot \xi) = \bar{\alpha} \circ \tilde{\xi}$ where $\bar{\alpha} \in [\Sigma S_R^m, Y_R]$ denotes the adjoint of α.

This pairing is not bilinear in general, it has the following properties:

(2. 12) Definition. Let L be a Lie algebra, see (0. 4). We say L is a Lie algebra over the homotopy coefficients $M = M_R^{**}$, if we have a pairing

$$L_m \times M_R^{m,n} \xrightarrow{\odot} L_n$$

which satisfies the following set of axioms: ($\alpha, \beta \in L$, $\xi, \eta \in M$)

(a) $\alpha \odot e^m = \alpha$

(b) $(\alpha \odot \xi) \odot \eta = \alpha \odot (\zeta \odot \eta)$

(c) $[\alpha \odot \xi, \beta \odot \eta] = \sum_{N, M \geq 1} [\alpha^M, \beta^N] \odot (\gamma_M(\zeta) \# \gamma_N(\eta))$ (compare II (3. 4)').

(d) The pairing is not bilinear, we have

$$\begin{cases}(\lambda\alpha)\odot\xi=\alpha\odot(\lambda\zeta)=\lambda(\alpha\odot\zeta) & \text{for } \lambda\in R\\ (\alpha\odot\zeta)+(\alpha\odot\eta)=\alpha\odot(\zeta+\eta) & \text{however}\\ (\alpha\odot\zeta)+(\beta\odot\zeta)=(\alpha+\beta)\odot\zeta+\sum_{n\geq 2} c_n(\alpha,\beta)\odot\gamma_n(\zeta)\end{cases}$$

where $c_n(\alpha, \beta)$ is the Zassenhaus term

$$c_n(\alpha,\beta)=\sum_{d\in D_n}[\alpha,\beta]_{\phi(d)}$$

evaluated in the Lie algebra L, see II (2. 8). Since it is only for even degrees of ζ that $\gamma_n(\zeta)$ may be non trivial we have no signs in the formula for $c_n(\alpha, \beta)$.

Let Lie_M be the category of Lie algebras with coefficients M. Morphisms are the homomorphisms $f : L \to L'$ in Lie_R with $f(\alpha \odot \zeta) = \zeta \odot f(\alpha)$ for $\zeta \in M$, $\alpha \in L$, see (0. 4).

Again we know from chapter II and III:

(2. 13) **Proposition.** With the structure (2. 11) the homotopy Lie algebra is a covariant functor

$$L(\,.\,, R) : \mathrm{Top}_1 \to \mathrm{Lie}_M$$

with $L(Y, R) = \pi_*(\Omega Y) \otimes R$, see (1. 1). For $f : Y \to Y'$ in Top_1 we have $f_*(\alpha) = (\Omega f) \circ \alpha$ for $\alpha \in L(Y, R)$.

§ 3. The Hurewicz and the degree map

Let $e_n \in H_n(S^n, \mathbb{Z}) \subset H_n(S^n, R)$ be an integral generator and let $e^n \in H^n(S^n, R)$ be the dual generator with $e^n(e_n) = 1$. We have natural transformations of functors

(3. 1) $\phi : \pi_*(\Omega Y) \otimes R \to H_*(\Omega Y, R)$

with $\phi(y) = y_*(e_n)$.

(3. 2) $\deg : M^*(X, R) \to H^*(X, R)$

with $\psi(x) = s^{-1}\tilde{x}^* s e^n$

V

where s denotes suspension, compare the definition of $\tilde{x}$ in (1. 6). ϕ is the Hurewicz map and deg is a variant of the classical degree map.

We now describe to what extent ϕ and deg preserve the algebraic structures. Clearly for the augmentation $\varepsilon : M_R^{**} \to R$ with $\varepsilon(e^m) = 1$ and $\varepsilon(\zeta) = 0$ for $\zeta \in M^{m,n}$ with $m \neq n$ we have

$$\begin{cases} \phi(\alpha \odot \zeta) = \phi(\alpha) \cdot \varepsilon(\zeta) \\ \deg(\zeta \odot x) = \varepsilon(\zeta) \cdot \deg(x) . \end{cases}$$

We now consider the Hurewicz map ϕ: The homology $H_*(G, R)$ of a connected loop space or topological group G is a graded (in general non commutative) algebra. The product is the Pontrjagin product

$$x * y = \mu_*(x \times y) \quad (x, y \in H_*(G, R))$$

where

(3. 3) $\quad \times : H_*(G, R) \otimes H_*(G, R) \to H_*(G \times G, R)$

is the cross product and where $\mu : G \times G \to G$ is the multiplication on G.

We associate with the product $*$ the bracket

$$[x, y] = x * y - (-1)^{|x|\,|y|} y * x,$$

so that $H_*(G, R)$ is a Lie algebra. It is a result of Samelson that ϕ in (3. 1) is a homomorphism of Lie algebras. Equivalently we may say that for the universal enveloping algebra U( ) the Hurewicz map ϕ gives us a map

(3. 4) $\quad \hat{\phi} : U(\pi_*(G) \otimes R) \to H_*(G, R)$

of algebras. For a Lie algebra L, $U(L)$ is not only an algebra, but a Hopf algebra with a commutative diagonal Δ and with primitive elements $PU(L) = L$, see [15].

(3. 5) **Proposition.** If the cross product (3. 3) is an isomorphism, the diagonal $\Delta : G \to G \times G$ induces a Hopf algebra structure on $H_*(G, R)$ and then $\hat{\phi}$ in (3. 4) is a homomorphism of Hopf algebras.

The cross product $\times$ is an isomorphism, if, for example, $H_*(G, R)$ is a free R-module, in particular, if $R = \mathbb{Q}$.

(3.6) Milnor-Moore theorem [33]. For $R = \mathbb{Q}$, $\hat{\phi}$ is an isomorphism of Hopf algebras.

We now consider the degree map in (3.2). The cohomology $H^*(X, R)$ is a graded commutative algebra. The product is the cup product. It is easily seen that deg is a homomorphism of commutative algebras. Moreover for the divided powers γ on $M^*(X, R)$ we have the equation

(3.7) $r\,!\ \deg(\gamma_r x) = (\deg(x))^r$

Therefore, if deg is surjective and if $H^*(X, R)$ is a free R-module, we see that the degree map is a homomorphism of algebras with divided powers. In particular, we have by (1.4):

(3.8) Proposition. For $R = \mathbb{Q}$ the degree map $\deg : M^*(X, \mathbb{Q}) \cong H^*(X, \mathbb{Q})$ is an isomorphism of algebras with divided powers.

To some extent this is the dual of the Milnor-Moore theorem.

Proof of (3.7). Serre [36] has shown that the integer cohomology of the loop space $\Omega\Sigma S^n$ of a sphere is an algebra with divided powers over $\mathbb{Z}$. $H^{nk}(\Omega\Sigma S^n, \mathbb{Z})$ is a free $\mathbb{Z}$-module generated by x_k and if n is even the cup product is determined by $x_1^k = k!\, x_k$. //

Proof of (3.8). We have to prove that for n odd

$$\deg : M^n(X, \mathbb{Q}) = [X, S^n_{\mathbb{Q}}] \xrightarrow{\cong} H^n(X, \mathbb{Q})$$

is an isomorphism of $\mathbb{Q}$-vector spaces. Clearly deg is a bijection by (1.4). In fact (1.4) is an equivalence of H-spaces. We show that the abelian group structure induced by deg on $[X, S^n_{\mathbb{Q}}]$ is the one defined in (ii) of (1.7). For this we consider the suspension map

$$\Sigma : [X, S^n_{\mathbb{Q}}] \to [\Sigma X, \Sigma S^n_{\mathbb{Q}}].$$

$\vee$
For the H-space $S^n_{\mathbb{Q}} = K(\mathbb{Q}, n)$ this map is not a group homomorphism since $K(\mathbb{Q}, n) \to \Omega\Sigma K(\mathbb{Q}, n)$ is not an H-map. From the Hopf construction on the multiplication of $K(\mathbb{Q}, n)$ we see that

$$-\Sigma x - \Sigma y + \Sigma(x+y) = [j_n, j_n]\,\Sigma\left(\frac{x \cup y}{2}\right) \text{ . } //$$

Clearly for $R \neq \mathbb{Q}$ the functor M^* is much more complicated than cohomology. Still, if $H^*(X, R)$ is finitely generated as an R-module then so is $M^*(X, R)$. Moreover we have the following special case.

(3.9) Proposition. Let X be a CW-complex of dimension k which is (c - 1)-connected, then deg in (3.2) is an isomorphism if $1/p \in R$ for all primes p with $p < (k - c + 3)/2$.

Proof. Since $\pi_{2p-3+n}(S^n)$ (n odd), is the lowest group in $\pi_*(S^n)$ containing p-torsion, see 9.7.13 [38], we know

$$S^n_R \xrightarrow{i} K(R, n) \qquad n \text{ odd}$$

$$\Omega\Sigma S^n_R \underset{\Omega i}{\hookrightarrow} \Omega K(R, n+1) = K(R, n) \qquad n \text{ even}$$

are k-connected if $1/p \in R$ for all primes p with $n + 2p - 3 \leq k - 1$. //

Proposition (3.9) is a generalization of (3.8), thus there should be a dual of (3.9) which generalizes the Milnor-Moore theorem (3.6).

We now consider the special situation where the Hurewicz map or the degree map is surjective. We say a space Y is of finite type over R, if $H_n(Y, R)$ is a finitely generated R-module for all n.

(3.10) Lemma. Let Y be a simply connected space of finite type over R. Then the following statements are equivalent:

(i) $H_*(\Omega Y, R)$ is a free R-module and $\hat{\phi} : U\pi_*(\Omega Y) \otimes R \to H_*(\Omega Y, R)$ is surjective, see (3.4).

(ii) There exists a homotopy equivalence

$$\Omega Y_R \simeq \underset{i=1}{\overset{N}{\times}} \Omega^{n_i}_R \text{ , see (1.3),}$$

with $n_1 \le n_2 \le \dots$ *and* $N \le \infty$.

For $N = \infty$ *the sequence* n_i *tends to* ∞ *and the product has the weak topology.*

If (i) or (ii) are satisfied we say $\Omega Y_{\mathbf{R}}$ is *decomposable.*

Proof of (3.10): (ii) $\Rightarrow$ (i). Let $\alpha_i : S_{\mathbf{R}}^{n_i} \to \Omega Y_{\mathbf{R}}$ be the restriction of the homotopy equivalence α in (ii) to $S_{\mathbf{R}}^{n_i} \subset \Omega_{\mathbf{R}}^{n_i}$. If n_i is even we extend α_i as an H-map to $J(S_{\mathbf{R}}^{n_i}) = \Omega_{\mathbf{R}}^{n_i}$. The product of these extensions is then again a homotopy equivalence β. In fact, β induces the same map in cohomology as α. This implies that ϕ is surjective.

(i) $\Rightarrow$ (ii). Since $\hat{\phi}$ is surjective the canonical map

$$UPH_*(\Omega Y_{\mathbf{R}}) \underset{\tau}{\overset{\cong}{\to}} H_*(\Omega Y_{\mathbf{R}})$$

is an isomorphism. Clearly τ is surjective since $\hat{\phi}$ factors over τ. Since $H_*(\Omega Y_{\mathbf{R}})$ is a free module, we have injective maps i, j in the commutative diagram

$$\begin{array}{ccc}
UPH_*(\Omega Y_{\mathbf{R}}) & \xrightarrow{\;\; i \;\;} & UPH_*(\Omega Y_{\mathbf{Q}}) \\
\tau \big\downarrow & & \cong \big\downarrow \tau \\
H_*(\Omega Y_{\mathbf{R}}) & \xrightarrow[\;\; j \;\;]{} & H_*(\Omega Y_{\mathbf{Q}})
\end{array}$$

Therefore by (3.6), τ is also injective. The isomorphism τ shows that $\hat{\not{c}} = \tau(U\not{c})$. Thus ϕ in (3.1) must also be surjective. We now choose an ordered base b_i, $i \ge 1$, of $PH_*(\Omega Y_{\mathbf{R}})$ and we choose α_i with $\phi(\alpha_i) = b_i$. These maps $\alpha_i : S_{\mathbf{R}}^{n_i} \to \Omega Y_{\mathbf{R}}$, $n_i = |b_i|$, yield a map

$$\beta : \times \Omega_{\mathbf{R}}^{n_i} \to \Omega Y_{\mathbf{R}}$$

as in the proof (ii) $\Rightarrow$ (i) above. Clearly β induces an isomorphism

$$\beta_* : PH_*(\times \Omega_{\mathbf{R}}^{n_i}) \overset{\cong}{\to} PH_*(\Omega Y_{\mathbf{R}})\,.$$

By use of the Poincaré-Birkhoff-Witt theorem (see 2.6 in [15]) the isomorphism τ shows that

V

$$\beta_* : H_*(\times \Omega_R^{n_i}) \cong H_*(\Omega Y_R)$$

is an isomorphism of R-modules. Thus β is a homotopy equivalence. //

More easily we obtain the dual of (3.10):

(3.11) Lemma. Let X be a connected space of finite type over R. Then the following statements are equivalent

(i) $H^*(X, R)$ is a free R-module and $\deg : M^*(X, R) \to H^*(X, R)$ is surjective, see (3.2).

(ii) There exists a homotopy equivalence

$$\Sigma X_R \simeq \bigvee_{i=1}^{N} S_R^{n_i}$$

with $n_1 \leq n_2 \leq \ldots$ and $N \leq \infty$. For $N = \infty$ the sequence n_i tends to ∞.

In this case we say ΣX_R is decomposable.

Proof. We prove (ii) $\Rightarrow$ (i). We choose a bases $B = \{b_1, b_2, \ldots\}$ of $\tilde{H}^*(X, R)$ and elements β_i with $\deg(\beta_i) = b_i$ and $b_i = |b_i| + 1$. Then the sum

$$\tilde{\beta} = \sum_i \tilde{\beta}_i$$

that is, the limit of the finite subsums, is a homotopy equivalence. //

(3.12) Definition. For the double graded R-module of coefficients $M = M_R^{**}$ in § 2 and for graded R-modules

$$H = \{H^n, n \geq 0\} \text{ and } \pi = \{\pi_n, n \geq 0\}$$

we define the tensor product $M \underline{\otimes} H$ and $\pi \otimes M$ as graded modules by

$$(M \underline{\otimes} H)^n = \times_{j \geq n} M_R^{n,j} \otimes H^j$$

$$(\pi \otimes M)_n = \bigoplus_{n \geq j} \pi_j \otimes M^{j,n} .$$

(3.13) Theorem. Let ΣX_R and ΩY_R be decomposable. Then there exist isomorphisms of R-modules

$$M^*(X, R) = M \underline{\otimes} H^*(X, R)$$

$$\pi_*(\Omega Y) \otimes R = PH_*(\Omega Y, R) \otimes M .$$

Proof of (3.13). We choose a basis $\{b_1, b_2, \dots\}$ of $PH_*(\Omega Y, R)$ and elements α_i with $\phi(\alpha_i) = b_i$ as in the proof of (3.10). Then it is an easy consequence of (3.10) and of (2.11) that

$$\alpha_* : (PH_*(\Omega Y, R) \otimes M)_n \xrightarrow{\cong} \pi_n(\Omega Y) \otimes R$$

$$\sum_i b_i \otimes \zeta_i \to \sum_i \alpha_i \odot \zeta_i$$

is an isomorphism of R-modules.

Now we choose a basis $B = \{b_1, b_2, \dots\}$ and elements β_i as in the proof of (3.11). This yields the isomorphism

$$\beta_* : (M \underline{\otimes} H^*(X, R))^n \xrightarrow{\cong} M^n(X, R)$$

$$\sum_i \zeta_i \otimes b_i \mapsto \sum_i \zeta_i \odot \beta_i \ .$$

If X is not finite dimensional, we use the fact that $M^n(X, R)$ is an inverse limit, see (1.5). Assume now, X is finite dimensional.

If n is even, it is an easy consequence of (3.11) that β_* is an isomorphism. If n is odd, we consider the diagram

$$\begin{array}{ccc}
M^n(X, R) = [X, S^n_R] & \xleftarrow{\ \beta_*\ } & \bigoplus_{b\in B} M_R^{n, |b|} \\
\Big\downarrow \Sigma & & \\
[\Sigma X, \Sigma S^n_R] & \overset{\widetilde{\beta}^*}{\approx} & \prod_{b\in B} \pi_{|b|+1}(\Sigma S^n_R) \\
\Big\downarrow \gamma_2 & & \\
[\Sigma X, \Sigma S^{2n}_R] & &
\end{array}$$

where suspension Σ is an inclusion and where image (Σ) = kernel (γ_2). From (3.11) we have the bijection $\widetilde{\beta}^*$.

We deduce from (ii) in (1.7) and (2.1) that $\Sigma\beta_*$ and thus β_* is injective. For surjectivity of β_* it is enough to prove $\ker \gamma_2 \subset \mathrm{im}(\Sigma\beta_*)$. It follows from (2.2) that any $\alpha \in [\Sigma X, \Sigma S^n_R]$ has a unique presentation

$$\alpha = \overset{V}{\sum_{b \in B}} ((\Sigma \xi_b) + [j, j] \circ \eta_b) \circ \widetilde{\beta_*(b)}$$

with $\xi_b \in M_R^{n, |b|}$, $\eta_b \in M_R^{2n, |b|}$.

From (ii) in (1. 7) and from the formulas in III we derive

$$\alpha = \Sigma \beta_*(\xi_b | b \in B) + [j, j]W$$

with an appropriate W. Thus $\gamma_2 o = 2W$, see III, and therefore $\gamma_2 \alpha = 0$ iff $W = 0$. //

There are various well-known examples of decomposable suspensions and loop spaces in literature:

(3. 14) **Lemma.** (A) If ΣX and ΣY are decomposable then also $\Sigma(X \times Y)$.

(B) If ΩX and ΩY are decomposable then also $\Omega(X \vee Y)$.

Proof. This is a consequence of the well known homotopy equivalences of D. Puppe and Ganea

$$\Sigma(X \times Y) \simeq \Sigma X \vee \Sigma Y \vee \Sigma X \wedge Y$$

and

$$\Omega(X \vee Y) \simeq \Omega X \times \Omega Y \times \Omega\Sigma(\Omega X \wedge \Omega Y),$$

see [11].//

(3. 15) **Examples.** The following loop spaces and suspensions are decomposable:

(a) $\Omega\Sigma S_R^n \simeq \Omega_R^n \times \Omega_R^{2n}$, n odd,

(b) $\Omega(\Sigma S^{n_1} \vee \ldots \vee \Sigma S^{n_k})$, more generally,

(c) let $T = \Sigma S^{n_1} \times \ldots \times \Sigma S^{n_k}$ be a product of spheres and let $T(n)$ be the union of all n-fold subproducts, then $\Omega T(n)$ is decomposable.

(d) Σ (product of spheres)

(e) $\Sigma(\Omega\Sigma S_R^n)$.

[(a) and (e) were proved by James, (b) is the original result of Hilton

[23] and follows from (3.14) as well as (d). Moreover (c) is a result of Porter, see: Am. J. Math. 87 (1965) 297-314, see also Am. J. Math. 88 (1966) 655-63.]

There are many more decomposable suspensions as shown by the following remark, which is proved in [9].

(3.16) **Remark.** For each rational space $X_{\mathbb{Q}}$ of finite type there is a space Y with $Y_{\mathbb{Q}} \simeq X_{\mathbb{Q}}$ so that ΣY_R is decomposable.

The statement dual to this remark should be true as well. Clearly any rational suspension or loop space of finite type is decomposable.

VI. GROUPS OF HOMOTOPY CLASSES

§ 1. Nilpotent rational groups of homotopy classes

We first consider the rationalization $[X, G]_{\mathbb{Q}}$ of the nilpotent group $[X, G]$. This study will serve as a guide in the more complicated situation $R \subset \mathbb{Q}$, $R \neq \mathbb{Q}$.

A group G or a Lie algebra L is nilpotent if there exists an integer $k \geq 1$ such that an iterated bracket of any k of its elements taken in any order is zero. The bracket is the commutator in G and the Lie product in L.

For a nilpotent (non graded) Lie algebra L the Baker-Campbell-Hausdorff formula (see chapter I)

$$x \cdot y = x + y + \tfrac{1}{2}[x, y] + \frac{1}{12}[[x, y], y] + \frac{1}{12}[[y, x], x] + \dots$$

provides a group multiplication on the underlying set of L. This group $\exp(L) = (L, \cdot)$ is a nilpotent rational group. Moreover the correspondence $L \mapsto \exp L$ is even an equivalence of categories as shown by Malcev.

The following types of Lie algebras appear naturally in homotopy theory.

(1.1) Definition. Let (C, Δ) be a graded commutative co-algebra over R with $C_0 = R$ and let $(\pi, [\ ,\ ])$ be a graded Lie algebra over R with $\pi_0 = 0$. The R-module of degree zero homomorphisms

$$\mathrm{Hom}_R(C, \pi)$$

is a non-graded Lie algebra over R with the Lie bracket defined by

$$[f, g] : C \xrightarrow{\Delta} C \otimes C \xrightarrow{f \otimes g} L \otimes L \xrightarrow{[\ ,\ ]} L$$

If $C_n = 0$ for all $n > N$ or if $L_n = 0$ for all $n > N$, then $\mathrm{Hom}_R(C, \pi)$ is a nilpotent Lie algebra over R. In I (3. 7) we gave a presentation of the group

$$\exp \mathrm{Hom}_{\mathbb{Q}}(C, \pi)$$

in terms of generators and relations.

If $\mathrm{Hom}_{\mathbb{Q}}(C, \pi)$ is not nilpotent we define the group

$$(1.2) \qquad \exp \mathrm{Hom}_{\mathbb{Q}}(C, \pi) = \varprojlim \exp \mathrm{Hom}_{\mathbb{Q}}(C_{[n]}, \pi)$$

to be the inverse limit of nilpotent groups. Here $C_{[n]}$ is the sub coalgebra of C of all elements of degree $\leq n$. The inverse limit is taken over the inclusions $C_{[1]} \subset C_{[2]} \subset \ldots \subset C$.

As an example of this group we take for C the homology coalgebra $H_*(X, \mathbb{Q})$ and for π the homotopy Lie algebra $\pi_*(\Omega Y) \otimes \mathbb{Q}$. We prove:

(1. 3) Theorem. Let X and ΩY be connected CW-spaces of finite type over $\mathbb{Q}$. Then we have an isomorphism σ of groups

$$[\Sigma X, Y]_{\mathbb{Q}} \cong \exp \mathrm{Hom}_{\mathbb{Q}}(H_*(X, \mathbb{Q}), \pi_*(\Omega Y) \otimes \mathbb{Q})$$

which is natural with respect to co-H-maps $\Sigma X \to \Sigma X'$ and H-maps $\Omega Y \to \Omega Y'$, see II (1. 3).

Since $\Sigma X_{\mathbb{Q}}$ is decomposable it follows easily that $[\Sigma X, Y]_{\mathbb{Q}} = [\Sigma X_{\mathbb{Q}}, Y_{\mathbb{Q}}]$ is the inverse limit of the groups $[\Sigma X^N, Y]_{\mathbb{Q}}$.

Remark. The isomorphism in (1. 3) is implied in the work of H. Scheerer [35]. The result can also be derived from Thom's and Sullivan's minimal model [40] of the function space Y^X. This model is constructed by S. Halperin and C. Watkiss in [22], (see also [20, 37]), and by use of [14] we obtain the fundamental group $\pi_1(Y^X)_{\mathbb{Q}} = [\Sigma X, Y]_{\mathbb{Q}}$ from it. Still, this very neat formula (1. 3) does not appear in the literature.

Our method of proving (1. 3) is different from those in the remark. The part of rational homotopy theory we need is that an odd dimensional sphere S^n is rationally an Filenberg-MacLane space $K(\mathbb{Q}, n)$, a result first proved by J. P. Serre in [36].

The isomorphism σ in theorem (1. 3) can be characterized by use of the elements $\tilde{x}$ in V (1. 6). In fact σ is the unique homomorphism of groups satisfying

$$(1.4) \qquad \sigma(x \otimes \alpha) = \alpha \circ \tilde{x} \text{ for}$$

$$\begin{cases} x \in M^n(X, \mathbb{Q}) = H^n(X, \mathbb{Q}) = \mathrm{Hom}(H_n(X, \mathbb{Q}), \mathbb{Q}) \\ \alpha \in \pi_n(\Omega Y) \otimes \mathbb{Q} = [\Sigma S^n_{\mathbb{Q}}, Y_{\mathbb{Q}}] \end{cases}$$

with $n \geq 1$. In (1. 4) the element $x \otimes \alpha \in \mathrm{Hom}(H_*, \pi_*)$ is the homomorphism $t \mapsto x(t) \cdot \alpha$ for $t \in H_*$. Moreover the composition $\alpha \circ \tilde{x} : \Sigma X_{\mathbb{Q}} \to \Sigma S^n_{\mathbb{Q}} \to Y_{\mathbb{Q}}$ is an element of $[\Sigma X, Y]_{\mathbb{Q}}$ by V (0. 2).

Since $\Sigma X_{\mathbb{Q}}$ or $\Omega Y_{\mathbb{Q}}$ is decomposable we see easily that the elements of the form $\alpha \circ \tilde{x}$ generate the group $[\Sigma X, Y]_{\mathbb{Q}}$.

§ 2. The exponential group

In § 1 we consider the exponential group structure on the rational Lie algebra of homomorphisms $\mathrm{Hom}_{\mathbb{Q}}(C, \pi)$.

We here generalize this type of group to the non rational case. Let R be a subring of $\mathbb{Q}$ with $1/2, 1/3 \in R$ and let $K = (K^{**}, \odot, \#, \gamma_r, e^r)$ be a module of homotopy coefficients, V §2.

Assume we have a K-algebra $A \in \mathrm{Div\ alg}_K$ and a Lie algebra π over K in Lie_K. These algebras are subject to pairings

$$K^{m,n} \times A^n \to A^m, \ (k, x) \mapsto k \odot x$$

$$\pi_m \times K^{m,n} \to \pi_n, \ (\alpha, k) \mapsto \alpha \odot k$$

as we defined in V (2. 9), V (2. 12).

If A is finitely generated we define the exponential group:

$$(2.1) \qquad \exp_K(A, \pi) = FG\Big(\bigcup_{n \geq 1} A^n \times \pi_n\Big)/\sim$$

as follows: FG(M) denotes the free group generated by the set M. The relation $\sim$ for the exponential group is generated by the relations (i) ... (iv):

(i) $(x, \alpha)^{-1}(y, \alpha)^{-1}(x+y, \alpha) \sim (\frac{x \cup y}{2}, [\alpha, \alpha])$

(ii) $(x, \alpha)(x, \beta) \sim (x, \alpha+\beta) \cdot \prod_{n\geq 2} (\gamma_n x, c_n(\alpha, \beta))$

(iii) $(x, \alpha)^{-1}(y, \beta)^{-1}(x, \alpha)(y, \beta) \sim \prod_{n\geq 1} \prod_{m\geq 1} (\gamma_m x \cup \gamma_n y, R_{m,n}(\alpha, \beta))$

(iv) $(k \odot x, \alpha) \sim (x, \alpha \odot k)$ for $k \in K^{|\alpha|, |x|}$

where $x, y \in A$ and $\alpha, \beta \in \pi$. Clearly in (i) we have $|x| = |y| = |\alpha|$ and in (ii) $|x| = |\alpha| = |\beta|$.

The products in (ii) and (iii) are finite since we replace $(0, \alpha)$ by the neutral element of the free group. The Zassenhaus term $c_n(\alpha, \beta)$ and the commutator term $R_{m,n}(\alpha, \beta)$ in (ii) and (iii) are given by the formulas in I (1.13) and I (2.6). Clearly these formulas are evaluated here in the Lie algebra π.

If A is not finitely generated but of finite type we define

(2.4) $\exp_K(A, \pi) = \lim\limits_{\leftarrow} \exp_K(A_{[n]}, \pi)$

where $A_{[n]}$ is the R-module obtained from A by dividing out all elements $\zeta \odot x$ with $\zeta \in M^{k,m}$, $x \in A^m$ and $m > n$, $k \geq 1$. Clearly $A^k_{[n]} = 0$ for $k > n$. The quotient map

$$q : A \to A_{[n]}$$

induces in a unique way the structure of an M-algebra on $A_{[n]}$. The inverse limit is taken over the projections $\to A_{[n]} \to A_{[n-1]} \to \dots$. Here we use the following functorial properties of the construction $\exp_K(A, \pi)$.

Clearly any triple

$$f : A \to A'$$
$$\chi : K \to K'$$
$$g : \pi \to \pi'$$

of homomorphisms with $f(k \odot x) = \chi(k) \odot f(x)$ and of $g(\alpha \odot k) = \chi(\alpha) \odot g(k)$ induces a homomorphism of groups

(2.5) $(f, \chi, g)_* : \exp_K(A, \pi) \to \exp_{K'}(A', \pi')$

mapping a generator (x, α) to $(fx, g\alpha)$. We say in this case f and g are equivariant with respect to χ.

§ 3. Groups of homotopy classes

Since for finite dimensional X or finite codimensional Y the group $[\Sigma X, Y]$ is nilpotent we can consider the R-localization $[\Sigma X, Y]_R$ of this group with respect to any subring $R \subset \mathbb{Q}$ of the rationals. We now show that $[\Sigma X, Y]_R$ is an exponential group in the sense of § 2 if ΣX or ΩY are decomposable. Moreover we consider the following problems which are dual to each other:

(A) Under what condition on X the group $[\Sigma X, Y]_R$ can be fully described in terms of the cohomology ring $H^*(X, R)$ and the homotopy Lie algebra $\pi_*(\Omega Y) \otimes R$?

(B) Under what condition on Y the group $[\Sigma X, Y]_R$ can be fully described in terms of the cohomotopy algebra $M^*(X, R)$ and the homology Lie algebra $PH_*(\Omega Y, R)$?

We know that for $R = \mathbb{Q}$ we need no condition on X or Y in (A) and (B) respectively. Clearly the case $R \neq \mathbb{Q}$ is more complicated.

Let X and ΩY be connected and assume X is finite dimensional or ΩY is finite codimensional. Then we introduce for the spherical cohomotopy algebra $M^*(X, R)$ and the homotopy Lie algebra $\pi_*(\Omega Y) \otimes R$ the homomorphism of R-local groups.

$$(3.1)\quad \exp_M(M^*(X, R), \pi_*(\Omega Y) \otimes R) \xrightarrow[\rho_R]{} [\Sigma X, Y]_R$$

where $M = M_R^{**}$ denotes the coefficients in V § 2. On generators (x, α) the homomorphism ρ_R is defined by

$$\rho_R(x, \alpha) = \alpha \circ \tilde{x}$$

where $\alpha \in \pi_n(\Omega Y) \otimes R = [\Sigma S^n_R, Y_R]$ and $\tilde{x} \in [\Sigma X_R, \Sigma S^n_R]$, see V (1. 6).

In chapter II we have proved that ρ_R is a well defined homomorphism. If $R = \mathbb{Q}$ then ρ_R in (3. 1) is exactly the isomorphism described in (1. 3), compare I (3. 7).

If ΣX or ΩY are decomposable, see V (3. 11) and V (3. 10), we will prove that ρ_R is also an isomorphism. In fact we even prove for the non finite dimensional case

(3.2) Theorem. Let X and ΩY be connected and of finite type over R. If ΣX or ΩY are decomposable there is an isomorphism of R-local groups

$$\exp_M(M^*(X, R), \pi_*(\Omega Y) \otimes R) \cong [\Sigma X_R, Y_R]$$

which is natural with respect to Co-H-maps, $\Sigma X \to \Sigma X'$ and H-maps $\Omega Y \to \Omega Y'$, see § 4.

(3.3) Definition. (A) Let ΣX be decomposable. Then the degree map

$$\deg : M^*(X, R) \to H^*(X, R)$$

is a surjective map of algebras with divided powers, see V (3.7). We say ΣX is splittable if there exists a right inverse

$$\sigma : H^*(X, R) \to M^*(X, R)$$

of deg ($\deg \sigma = \mathrm{id}$) which is also a homomorphism of algebras with divided powers.

(B) Let ΩY be decomposable. Then the Hurewicz map

$$\phi : \pi_*(\Omega Y) \otimes R \to PH_*(\Omega Y, R)$$

is a surjective map of Lie algebras. We say ΩY is splittable if there exists a right inverse

$$\tau : PH_*(\Omega Y, R) \to \pi_*(\Omega Y) \otimes R$$

of ϕ which is a homomorphism of Lie algebras.

(3.4) Theorem. Let X and ΩY be connected and of finite type over R. If ΣX is splittable, the homomorphism

$$\rho_R(\sigma, i, \mathrm{id})_* : \exp_{\underline{R}}(H^*(X, R), \pi_*(\Omega Y) \otimes R) \cong [\Sigma X, Y]_R$$

is an isomorphism of R-local groups. If ΩY is splittable, the homomorphism

$$\rho_R(\text{id}, i, \tau)_* : \exp_{\underline{R}}(M^*(X, R), PH_*(\Omega Y, R)) \cong [\Sigma X, Y]_R$$

is an isomorphism of R-local groups.

We use (2. 5) for the inclusion $i : \underline{R} \hookrightarrow M$, see V (2. 6). The theorem implies that in case ΣX and ΩY are splittable the group $[\Sigma X, Y]_R$ depends only on the cohomology ring $H^*(X, R)$ and on the homology Lie algebra $PH_*(\Omega Y, R)$.

The reason is that $M^*(X, R)$ for splittable ΣX is the 'M-extension of $H^*(X, R)$' and $\pi_*(\Omega Y) \otimes R$ for splittable ΩY is the M-extension of $PH_*(\Omega Y, R)$, see chapter VII.

Proof of (3. 2) and (3. 4). We only prove (3. 2). With modifications we obtain along the same lines the proof of (3. 4). Let $G = [\Sigma X_R, Y_R]$ and let $G_{[n]}$ be the quotient group of G obtained by dividing out all elements

$$\Sigma X_R \to \Sigma S^q_R \to Y_R \quad \text{for } q \geq n.$$

If ΣX or ΩY is decomposable, it is easily seen by V (3. 11) and V (3. 10) that the canonical map

$$G \to \lim_{\leftarrow} G_{[n]}$$

is an isomorphism. With definition (2. 4), for the proof of (3. 2) it is enough to prove

$$(1) \qquad G_{[n]} \underset{\rho_R}{\cong} \exp_M(M^*(X, R)_{[n]}, \pi_*(\Omega Y) \otimes R)$$

where ρ_R is defined as in (3. 1) by $\rho_R(x, \alpha) = \alpha \circ \tilde{x}$. Using again V (3. 11) or V (3. 10) we see that ρ_R in (1) is in fact surjective. To prove injectivity we first observe that all cosets of

$$H = \exp_M(M^*(X, R)_{[n]}, \pi_*(\Omega Y) \otimes R)$$

are represented by words

$$(2) \qquad (x_1, \alpha_1) \dots (x_r, \alpha_r)$$

of generators (x_i, α_i), $x_i \in M^*(X, R)$ with $|x_i| \le n$. This we know since by (i) and (iv) in (2. 3) we have

(3) $$(0, \alpha) \sim (x, 0) \sim \text{neutral element}$$
$$(x, \alpha)^{-1} \sim (-x, \alpha) \sim (x, -\alpha)$$

We say a coset $g \in H$ has <u>length</u> $L(g) \le r$ if it contains a word of generators as in (2) of length r. For the assumption

(*) ΣX is decomposable and

(*)' ΩY is decomposable

respectively we prove the proposition of (3. 2) in parallel. First we choose as in the proof of V (3. 19) and V (3. 10)

(4) a basis $B = \{b_1, b_2 \dots\}$ of $\tilde{H}^*(X, R)$ with $|b_1| \le |b_2| \le \dots$ and elements $b_i \in M^*(X, R)$ with $\deg \underline{b}_i = b_i$.

(4)' a basis $B = \{\beta_1, \beta_2, \dots\}$ of $PH^*(\Omega Y, R)$ with $|\beta_1| \le |\beta_2| \le \dots$ and elements $\underline{\beta}_i \in \pi_*(\Omega Y) \otimes R$ with $\phi(\underline{\beta}_i) = \beta_i$.

Inductively we prove: Each element $g \in H$ contains an element of the form

(5) $$\prod_{i=1}^{N} (\underline{b}_i, \zeta_i) \text{ with appropriate } \zeta_i$$

(5)' $$\prod_{i=1}^{M} (x_i, \underline{\beta}_i) \text{ with appropriate } x_i$$

if the assumptions (*) and (*)' hold respectively. Here N and M are determined by

$$|b_N| = |\beta_M| = n \text{ but}$$
$$|b_{N+1}| = |\beta_{M+1}| = n + 1.$$

First we obtain (5) and (5)' for $L(g) = 1$. For a generator $(x, \zeta) \in H$ we know by use of (3. 13) that there exist elements x_i and ζ_i in M with

$$(6) \qquad x = \sum_{i=1}^{N} x_i \odot \underline{b}_i \quad \text{(if (*) holds).}$$

$$(6)' \qquad \zeta = \sum_{i=1}^{M} \underline{\beta}_i \odot \zeta_i \quad \text{(if (*)' holds).}$$

We call x_i and ζ_i the coordinates of x and ζ respectively. If we apply inductively the relations (i)... (iv) in (2. 3) we see that

$$(7) \qquad (x, \zeta) = (\sum_{i=1}^{N} x_i \odot \underline{b}_i, \zeta)$$

$$\sim \prod_{i=1}^{N} (\underline{b}_i, (x, \zeta)_i)$$

with appropriate $(x, \zeta)_i \in \pi_*(\Omega Y) \otimes R$. In fact

$$(x, \zeta)_i = \zeta \odot x_i + \sum_{j>k} \tfrac{1}{2}[\zeta, \zeta] \odot (x_j \# x_k) \odot (\underline{b}_j \cup \underline{b}_k)_i \,.$$

We call such elements multiplicative coordinates of (x, ζ) with respect to B. Similarly we get

$$(7)' \qquad (x, \zeta) = (x, \sum_{i=1}^{M} \underline{\beta}_i \odot \zeta_i)$$

$$\sim \prod_{i=1}^{M} ((x, \zeta)^i, \underline{\beta}_i)$$

with appropriate $(x, \zeta)^i \in M^*(X, R)$. We now define the connectivity of an element (x, ζ) by

$$(8) \qquad \theta(x, \zeta) \geq n \text{ if } x_i = 0 \text{ for } |\underline{b}_i| < n$$

$$(9) \qquad \theta(x, \zeta) \geq n \text{ if } \zeta_i = 0 \text{ for } |\underline{\beta}_i| < n.$$

Clearly for all (x, ζ) we have $\theta(x, \zeta) \geq 1$. Moreover we check that for the multiplicative coordinates we have

$$(x, \zeta)^i = 0 \text{ or } (x, \zeta)_i = 0$$

if $i < \theta(x, \zeta)$. This follows since $x_i = 0$ for $|b_i| < |x|$.

We now apply the same inductive process as in the proof of II (5. 9). This way we get the proposition in (5) and (5)'. As in the proof of II (5.9) we derive from (5) and (5)' injectivity of ρ_R. //

(3.5) **Examples of splittable suspensions and loop spaces.**

(a) The suspension $\Sigma\Omega\Sigma S^n$ is splittable

(b) The suspension of a product of spheres $\Sigma(\times_{i=1}^{N} S^{n_i})$ is splittable.

(c) More generally $\Sigma(\times_{i=1}^{N} \Omega_R^{n_i})$ is splittable, see V (1.3).

(d) Assume $\pi_i(X) \otimes R = 0$ for $i = 1, \dots, c-1$ and $H_n(X, R) = 0$ for $n > 2c$. If ΣX is decomposable then ΣX is also splittable. (Thus for example orientable surfaces are splittable over R.) This follows since by use of the Hopf classification theorem 4.3.14 [11] $\deg : M^n(X, R) \to H^n(X, R)$ is an isomorphism for the top dimension $n = 2c$.

(e) Proposition V (3.9) yields examples of splittable suspensions.

(f) The loop space of a wedge of spheres $\Sigma S^{n_1} \vee \dots \vee \Sigma S^{n_k}$ or of a fat wedge $T(n)$ is splittable, see V (3.15).

(g) See III (4.5).

§ 4. H-maps and Co-H-maps

In this section we consider the naturality of the isomorphisms obtained in § 3. First we observe (see II (1.3)):

(4.1) **Proposition.** Let X and X' be connected spaces. Any Co-H-map $f : \Sigma X \to \Sigma X'$ induces a homomorphism $f^* : M^*(X', R) \to M^*(X, R)$ of algebras with divided powers which is equivariant with respect to the coefficients M in V § 2. The homomorphism f^* is defined by

$$(f^*(x))^{\sim} = \tilde{x} \circ f$$

compare V (1.6).

Proof. From III (6.1) and II (2.7) it follows that $\gamma_2(\tilde{x} \circ f) = 0$ if $|x|$ is odd. Therefore $\tilde{x} \circ f$ is desuspendable. //

(4.2) **Proposition.** Let ΩY and $\Omega Y'$ be connected. An H-map

$g : \Omega Y \to \Omega Y'$ induces a map of Lie algebras $g_* : \pi_*(\Omega Y) \otimes R \to \pi_*(\Omega Y') \otimes R$ which is equivariant with respect to the coefficients M.

Proof. That $g_*(\alpha \odot \xi) = g_*(\alpha) \odot \xi$ follows from the fact that, for the H-map g, $(\Omega R) \circ (\Omega \Sigma g) \simeq g \circ (\Omega R)$ where R is the evaluation. //

(4. 3) Remark. If the homology Lie algebra is defined, it is clear that an H-map g induces a homomorphism of homology Lie algebras

$$g_* : PH_*(\Omega Y, R) \to PH_*(\Omega Y', R) .$$

It is not so clear under what condition on X, X' and R a Co-H-map $f : \Sigma X \to \Sigma X'$ induces a homomorphism of cohomology algebras $s^{-1} f^* s : H^*(X', R) \to H^*(X, R)$. If ΣX and $\Sigma X'$ are decomposable it follows from (4. 1) that $s^{-1} f^* s$ is in fact an algebra homomorphism.

From (4. 1) and (4. 2) we obtain

$$M : [\Sigma X, \Sigma X']^{\text{Co-H}} \to \text{Div alg}_M(M^*(X', R), M^*(X, R))$$
$$L : [\Omega Y, \Omega Y']^{H} \to \text{Lie}_M(\pi_*(\Omega Y) \otimes R, \pi_*(\Omega Y') \otimes R)$$

associating to a homotopy class of a Co-H-map f or of an H-map g the induced map which is equivariant with respect to the coefficients $M = M_R^{**}$.

(4. 4) Theorem. Let ΣX, $\Sigma X'$ and ΩY, $\Omega Y'$ be R-local and decomposable. Then M and L above are bijections of sets.

Proof. We define inverses M', L' of the functions M and L above as follows. Let $\rho = \rho_R$ be the isomorphism in (5. 3),

$$M'(\phi) = \rho^{-1}(\exp(\phi, 1)(\rho\, 1_{\Sigma X'})),$$
$$L'(\psi) = \rho^{-1}(\exp(1, \psi)(\rho\, 1_{\Omega Y})). \quad //$$

As a special case we obtain for $R = \mathbb{Q}$

(4. 5) Corollary. Let X, X', ΩY, $\Omega Y'$ be connected rational spaces of finite type. Then we have bijections of sets:

$$[\Sigma X, \Sigma X']^{\text{Co-H}} \approx \text{alg}_{\mathbb{Q}}(H^*(X', \mathbb{Q}), H^*(X, \mathbb{Q}))$$

where $alg_{\mathbb{Q}}$ denotes the set of algebra homomorphisms over $\mathbb{Q}$, and

$$[\Omega Y, \Omega Y']^H \approx \mathrm{Lie}_{\mathbb{Q}}(\pi_*(\Omega Y) \otimes \mathbb{Q}, \pi_*(\Omega Y') \otimes \mathbb{Q})$$

where $\mathrm{Lie}_{\mathbb{Q}}$ denotes the set of Lie algebra homomorphisms over $\mathbb{Q}$.

(4.6) Corollary. Let X, X', ΩY, $\Omega Y'$ be connected rational spaces of finite type.

(A) There is an equivalence $\Sigma X \sim \Sigma X'$ of Co-H-spaces if and only if there is an isomorphism of algebras $H^*(X, \mathbb{Q}) \cong H^*(X', \mathbb{Q})$.

(B) There is an equivalence $\Omega Y \sim \Omega Y'$ of H-spaces if and only if there is an isomorphism of Lie algebras $\pi_*(\Omega Y) \otimes \mathbb{Q} \cong \pi_*(\Omega Y') \otimes \mathbb{Q}$.

Clearly, corollary (4.6) has a generalization for decomposable suspensions or loop spaces which are R-local, $R \subset \mathbb{Q}$. In rational homotopy theory we have formal and coformal spaces:

(4.7) Definition. We say $\hat{X}$ is the formal type of a rational space X if the cohomology algebra $H^*(X, \mathbb{Q})$ is a Sullivan model for $\hat{X}$. We say $\check{Y}$ is the coformal type of the rational space Y if the homotopy Lie-algebra $\pi_*(\Omega Y) \otimes \mathbb{Q}$ is a Quillen model for $\check{Y}$, see [10].

(4.8) Corollary. Let X and ΩY be connected rational spaces of finite type. Then there are natural Co-H- and H-equivalences

$$\Sigma X \sim \Sigma \hat{X}, \quad \Omega Y \sim \Omega \check{Y}$$

respectively.

The question of realizing an abstract homomorphism $H^*(X, \mathbb{Q}) \to H^*(Y, \mathbb{Q})$ between cohomology algebras of given rational spaces X and Y is intensively studied in [17, 21, 42]. From the bijection in (4.5) we see that this problem is equivalent to the problem whether a given Co-H-map $\Sigma Y \to \Sigma X$ is actually a suspended map. This fact yields many examples of Co-H-maps which are not desuspendable.

The desuspension problem is naturally embedded in the problem of determining the James filtration:

(4.9) Definition. A homomorphism $\phi : H^*(X, \mathbb{Q}) \to H^*(Y, \mathbb{Q})$ of algebras has James filtration $\leq n$ with respect to rational spaces X and Y if the Co-H-map $M^{-1}(\phi) : \Sigma Y \to \Sigma X$ has James filtration $\leq n$. That is, the adjoint $\bar{\phi} : Y \to \Omega\Sigma X \simeq J(X)$ of $M^{-1}(\phi)$ factors over the inclusion $J_n(X) \subset J(X)$ of the n-fold reduced product space $J_n(X)$ see [27].

Clearly ϕ has James filtration 1 with respect to X and Y if and only if ϕ is realizable, that is if a map $f : Y \to X$ exists with $f^* = \phi$.

These results on rational spaces can partially be extended to splittable spaces.

Let $\Sigma\,\mathrm{Split}_R$ be the following category: Objects are pairs (X, σ) where ΣX is splittable and σ is a splitting of the degree homomorphism, see (3.3). Morphisms are homotopy classes of Co-H-maps $f : \Sigma X \to \Sigma X'$ which are compatible with the splittings that is $f^*\sigma' = \sigma f^*$.

Similarly let $\Omega\,\mathrm{Split}_R$ be the category of pairs (Y, σ) where ΩY is splittable and σ is a splitting of the Hurewicz homomorphism, see (3.3). Morphisms are homotopy classes of H-maps $g : \Omega Y \to \Omega Y'$ with $g_*\sigma = \sigma g^*$.

Using (4.3) we have functors

$$(4.10)\quad \begin{aligned} H^*(\ .\ , R) &: \Sigma\,\mathrm{Split}_R \to \mathcal{F}(\mathrm{Div\ alg}_R^*) \\ PH^*(\Omega\ .\ , R) &: \Omega\,\mathrm{Split}_R \to \mathcal{F}(\mathrm{Lie}_R) \end{aligned}$$

to the subcategories of $\mathrm{Div\ alg}_R^*$ and Lie_R of objects which are free R-modules of finite type. More generally than (4.5), we obtain from (3.4):

(4.11) Theorem. The functors (4.10) are bijective on morphism sets.

(4.12) Conjecture. The functors (4.10) are equivalences of categories.

It might be possible to prove this conjecture for $\Sigma\,\mathrm{Split}_R$ with a construction similar to the one of formal spaces in [17].

Clearly the conjecture is true for $R = \mathbb{Q}$.

VII. THE HILTON-MILNOR THEOREM AND ITS DUAL

§ 1. The category of coefficients

We here show that a module of homotopy coefficients as defined in V (2. 3) is equivalent to a monoid in the category 'coef_R' of coefficients.

Let $R \subset \mathbb{Q}$ be a subring of $\mathbb{Q}$.

(1. 1) Definition. We call a tuple $(M, \#, \gamma_p, e^r)$ a module of coefficients if

(i) $M = M_R^{**} = \{M^{m,n}, \ m, n \geq 1\}$ is a double graded R-module with

$$M^{m,n} = \begin{cases} 0 & m > n \\ R & m = n \\ \text{finite group} & m < n \end{cases}$$

(ii) $\# : M^{m,n} \otimes_R M^{m',n'} \to M^{m+m',n+n'}$ is a R-bilinear pairing $\zeta \otimes \eta \mapsto \zeta \# \eta$ for $m, n, m', n' \geq 1$.

(iii) $\gamma_p : M^{m,n} \to M^{mp,n}$, p a prime ≥ 2, is a homomorphism of R-modules $(m, n \geq 1)$.

(iv) $e^n = 1 \in R = M^{n,n}$, $n \geq 1$.

Moreover the following relations shall be satisfied.

(a) $\#$ is associative

(b) $\#$ is commutative in the sense $\zeta \# \eta = (-1)^{mm'+nn'} \eta \# \zeta$

(c) $e^n \# e^m = e^{n+m}$

(d) For e^1 the homomorphism

$$e : M^{m,n} \to M^{m+1,n+1}, \ e(\zeta) = e^1 \# \zeta$$

is an isomorphism for $n < 2m$. Moreover e maps an element of prime power order p^ν to an element of order p^ν or $p^{\nu-1}$.

(e) For each prime p we have

$$\gamma_p : M^{m,n} \to M^{pm,n}$$

with $p\gamma_p(\zeta) = 0$ and $\gamma_p(\zeta) = 0$ if m is odd. Moreover $\gamma_p(\zeta \# \eta) = 0$ for all $\zeta, \eta \in M$.

Let $coef_R$ be the category of coefficients over R. Morphisms are the maps $\phi : M \to N$ of bidegree $(0, 0)$ which are compatible with $\#$, γ, and e^r, that is

$$\begin{cases} \phi(\zeta \# \eta) = (\phi\zeta) \# (\phi\eta), & \\ \phi(\gamma_p \zeta) = \gamma_p(\phi\zeta), & p \text{ a prime,} \\ \phi(e^n) = e^n, & n \geq 1. \end{cases}$$

We now define a tensor product $\otimes$ in this category $coef_R$.

(1.2) **Definition.** Let M, N be modules of coefficients. Then we obtain the module $M \tilde{\otimes} N$ of coefficients as follows:

(1) $M \tilde{\otimes} N$ is the double graded R-module with

$$(M \tilde{\otimes} N)^{m,n} = \bigoplus_{m \leq j \leq n} M^{m,j} \otimes_R M^{j,n}$$

Then $M \tilde{\otimes} N$ as a module is generated by elements $\zeta \otimes \eta$ with $\zeta \in M^{m,j}$, $\eta \in M^{j,n}$.

(2) The bilinear pairing $\#$ on $M \tilde{\otimes} N$ is defined by

$$(\zeta \otimes \eta) \# (\zeta' \otimes \eta') = (\zeta \# \zeta') \otimes (\eta \# \eta').$$

(3) The invariants $\gamma_p : (M \tilde{\otimes} N)^{m,n} \to (M \tilde{\otimes} N)^{pm,n}$ are defined by

$$\gamma_p(\zeta \otimes \eta) = \gamma_p(\zeta) \otimes \eta + \zeta^{\#p} \otimes \gamma_p(\eta)$$

where $\zeta^{\#p} = \zeta \# \ldots \# \zeta$ is the p-fold product.

(4) $e^n \in (M \tilde{\otimes} N)^{n,n}$ is defined by $e^n = e^n \otimes e^n$.

(1.3) **Proposition.** If $1/2 \in R$ the structure $(\#, \gamma_p, e^n)$ on $M \tilde{\otimes} N$ above satisfies all relations $(a, \ldots, e)$ in (1.1).

Proof. We only check that γ_p is trivial on $(M \tilde{\otimes} N)^{m,n}$ if m is odd. For $\zeta \in M^{m,j}$, $\eta \in N^{j,n}$ with m and j odd we clearly have

$\gamma_p(\zeta \otimes \eta) = 0$. If j is even we have

$$\gamma_p(\zeta \otimes \eta) = \zeta^{\# p} \otimes \gamma_p(\eta).$$

However

$$\zeta \# \zeta = (-1)^{m^2+j^2} \zeta \# \zeta = -\zeta \# \zeta$$

Since $1/2 \in R$, $\zeta \# \zeta = 0$. //

In the following let $1/2 \in R$; then the tensor product $M \tilde{\otimes} N$ in Coef_R is well defined.

(1.4) **Proposition.** $\tilde{\otimes}$ is associative in Coef_R, that is, for coefficients M, N, P there is a canonical isomorphism

$$(M \tilde{\otimes} N) \tilde{\otimes} P = M \tilde{\otimes} (N \tilde{\otimes} P)$$

in Coef_R.

We have the trivial coefficients $\underline{R}$ in Coef_R with

(1.5) $$\underline{R}^{m,n} = \begin{cases} 0 & m \neq n \\ R & m = n \end{cases}$$

Clearly we always have the retraction

$$\underline{R} \overset{i}{\hookrightarrow} M \overset{\varepsilon}{\to} \underline{R}$$

in Coef_R. Moreover we have the canonical isomorphisms

(1.6) $$\underline{R} \tilde{\otimes} M = M = M \tilde{\otimes} \underline{R}.$$

These observations allow the following definition.

(1.7) **Definition.** A monoid in Coef_R is a module of coefficients M together with a pairing

$$\odot : M \tilde{\otimes} M \to M$$

which is associative and has the unit $\underline{R} \subset M$, this means, the diagrams

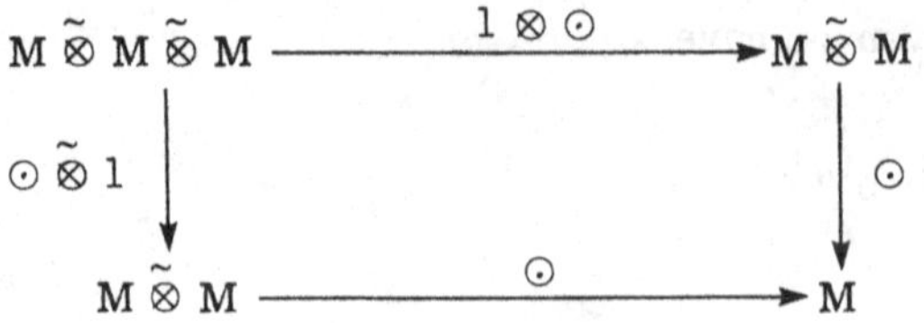

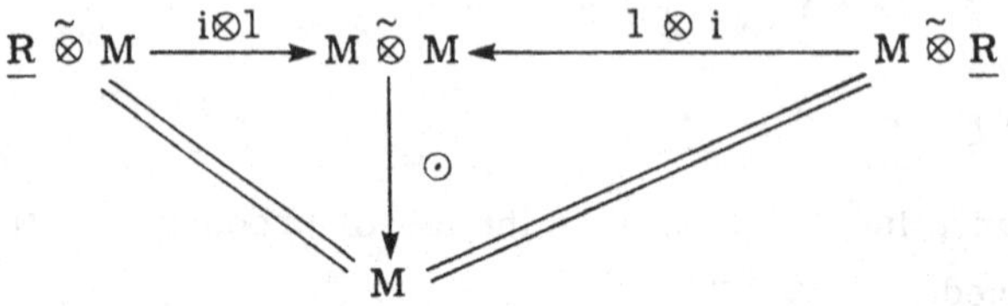

commute in Coef_R.

We now are ready to give an alternative definition of a module of homotopy coefficients, see V (2. 3):

(1. 8) **Proposition.** For $1/2 \in R \subset \mathbb{Q}$ modules of homotopy coefficients $(M, \odot, \#, \gamma, e)$ are in 1-1 correspondence with monoids $(M, \odot)$ in Coef_R.

Proof. We only check for a module of homotopy coefficients $M = (M, \odot, \#, \gamma, e)$ that

$$(1.9) \quad (\zeta \# \zeta') \odot (\eta \# \eta') = (\zeta \odot \eta) \# (\zeta' \odot \eta')$$

for $\zeta \otimes \eta \in M^{m,j} \otimes M^{j,n}$ and $\zeta' \otimes \eta' \in M^{m',j'} \otimes M^{j',n'}$. In fact by the relations (d) and (f) and (b) in V (2. 3) we have:

$$(\zeta \# \zeta') \odot (\eta \# \eta') = (\zeta \# e^{m'}) \odot (e^j \# \zeta') \odot (\eta \# e^{j'}) \odot (e^n \# \eta')$$
$$= (\zeta \# e^{m'}) \odot (\eta \# e^{m'}) \odot (e^n \# \zeta') \odot (e^n \# \eta')$$
$$= ((\zeta \odot \eta) \# e^{m'}) \odot (e^n \# (\zeta' \odot \eta'))$$
$$= (\zeta \odot \eta) \# (\zeta' \odot \eta').$$

This shows that for the coefficients $N = (M, \#, \gamma_p, e^n)$ given by M the operation $\odot$ in M yields a monoid structure on N. For this we observe that formula (i) in V (2. 3) reduces for a prime p because of (g) in V (2. 3) to

(1.10) $\gamma_p(\zeta \odot \eta) = (\gamma_p \zeta) \odot \eta + \zeta^{\# p} \odot \gamma_p(\eta)$.

Conversely if a monoid $(N, \odot)$ in Coef_R is given with $N = (M, \#, \gamma_p, e^n)$ we obtain the homotopy coefficients $(M, \odot, \#, \gamma_r, e^n)$. Here $\odot$, $\#$, e^n are the same as in N and the invariants

$$\gamma_r : M^{m,n} \to M^{rm,n}$$

are defined for a prime power $r = p^\nu$ by

(1.11) $\gamma_{p^\nu}(\zeta) = \gamma_p^\nu(\zeta)$

where $\gamma_p^\nu = \gamma_p \dots \gamma_p$ is the ν-fold composition. (1.11) follows from

$$1 \equiv \frac{[(p^{\nu-1}!)(p^{\nu-2}!)\dots(p^\nu!)]^p (p!)^\nu}{(p^\nu!)(p^{\nu-1}!)\dots(p^1!)}$$

mod p. //

§ 2. Extensions of algebras by homotopy coefficients

In V § 0 we defined the category $\mathrm{Div\,alg}_R$ of algebras with divided powers. We show in this section that modules of coefficients M in Coef_R operate on this category. This leads to an alternative definition of the category $\mathrm{Div\,alg}_M$ of M-algebras, defined in V (2.9). We show that for each algebra A in $\mathrm{Div\,alg}_R$ there exists an unique M-extension of A which is an M-algebra. In particular the M-extension of a free object in $\mathrm{Div\,alg}_R$ is a free object in $\mathrm{Div\,alg}_M$, so free objects exist in $\mathrm{Div\,alg}_M$.

Let R be a subring of $\mathbb{Q}$ with $1/2 \in R$.

(2.1) Definition. For an algebra $A = (A^*, \cup, \gamma)$ in $\mathrm{Div\,alg}_R$ and for a module of coefficients $M = (M^{**}, \#, \gamma, e^n)$ in Coef_R we define the twisted product $M \tilde{\otimes} A$, which is again an algebra with divided powers in $\mathrm{Div\,alg}_R$.

We first consider the case where A is finite dimensional, that is, there is $d \geq 0$, so that $A^n = 0$ for $n > d$. Then we define $M \tilde{\otimes} A$ as an R-module by $(M \tilde{\otimes} A)^0 = R$ and for $n \geq 1$ by

(1) $(M \tilde{\otimes} A)^n = \bigoplus_{j \geq n} M^{n,j} \otimes A^j.$

We define the structure of $M \tilde{\otimes} A$ as an object in Div alg_R by

(2) $(\zeta \otimes a) \cup (\eta \otimes b) = (\zeta \# \eta) \otimes (a \cup b)$

(3) $\gamma_n(\zeta \otimes a) = \sum_{r \geq 1} \Gamma_n^r(\zeta) \otimes \gamma_r(a)$

where γ, $\eta \in M$, $a, b \in A$. $\Gamma_n^r(\zeta)$ is defined in V (2. 3) (i). We extend $\cup$ bilinearly over $M \tilde{\otimes} A$ and we extend γ_n by formula V (0. 5) over $M \tilde{\otimes} A$, compare the proof of (2. 2) below.

If A is not finite dimensional we set

$$A^k_{[n]} = \begin{cases} 0 & k > n \\ A^k & k \leq n \end{cases}$$

The quotient map $A \to A_{[n]}$ gives $A_{[n]}$ the structure of an algebra with divided powers in Div alg_R. Now we define

(4) $M \tilde{\otimes} A = \lim\limits_{\leftarrow} M \otimes A_{[n]}$

to be the inverse limit in Div alg_R. As a module we have

$$(M \tilde{\otimes} A)^n = \times_{j \geq n} M^{n,j} \otimes A^j.$$

(2. 2) Proposition. By (2) and (3) $M \tilde{\otimes} A$ has a well-defined structure of an algebra with divided powers.

Proof. We consider the quotient map

(1) $U = \text{FAG}(\bigcup_{j \geq n} M^{n,j} \times A^j) \xrightarrow{\pi} \bigoplus_{j \geq n} M^{n,j} \otimes A^j$

where FAG(X) denotes the free abelian group generated by X. The relations of π are generated by

(2) $$\begin{cases} (\zeta, a) + (\zeta, b) \sim (\zeta, a + b) \\ (\zeta, a) + (\eta, a) \sim (\zeta + \eta, a) \\ (\lambda\zeta, a) \sim (\zeta, \lambda a) \quad \text{for } \lambda \in R \end{cases}$$

We now define for $r \geq 2$

(3) $\hat{\gamma}_r : U \to (M \tilde{\otimes} A)^{nr}$

by $\hat{\gamma}_r = 0$ if n is odd and if n is even by

(4) $\hat{\gamma}_r(\lambda(\zeta, a)) = \lambda^r \sum_{s \geq 1} \Gamma_r^s(\zeta) \otimes \gamma_s(a), \quad \lambda \in \mathbb{Z},$

and we extend over U by the formula $(u, v \in U)$

(5) $\hat{\gamma}_r(u + v) = \sum_{\substack{i+j=r \\ i, j \geq 0}} \hat{\gamma}_i(u) \cup \hat{\gamma}_j(v) .$

Here $\hat{\gamma}_0(u) = 1 \in (M \tilde{\otimes} A)^0$ and $\cup$ is the product which is well defined by the formula (2) in (2. 1). The cup product in (5) is symmetric, since $\hat{\gamma}_i(u) \in (M \tilde{\otimes} A)^{ni}$ where n is even. Therefore, there is in fact an unique map $\hat{\gamma}_r$ in (3) satisfying (4) and (5).

We have to show that $\hat{\gamma}_r$ factors as a function over π in (1). We first can check that for the relations in (2) we have for all r

$$\hat{\gamma}_r((\zeta, a) + (\zeta, b)) = \hat{\gamma}_r(\zeta, a + b)$$

$$\hat{\gamma}_r((\zeta, a) + (\eta, a)) = \hat{\gamma}_r(\zeta + \eta, a)$$

$$\hat{\gamma}_r((\lambda\zeta, a)) = \hat{\gamma}_r((\zeta, \lambda a))$$

Then $\hat{\gamma}_r$ factors over π since we derive from (5) for $u, v \in U$ with $\hat{\gamma}_r(u) = \hat{\gamma}_r(v)$ for $r \geq 2$ then also $\hat{\gamma}_n(x + u) = \hat{\gamma}_n(x + v)$ for all $x \in U$ and $n \geq 2$. //

(2. 3) **Remark.** The proof of (2. 2) is simpler if $R = \mathbb{Z}_{(p)}$ is the ring of p-local integers. Then we have for γ_i on $M \tilde{\otimes} A$ the formulas

$$\gamma_p(\zeta \otimes a) = \gamma_p(\zeta) \otimes a + \zeta^{\#p} \otimes \gamma_p(a)$$

and

$$\gamma_i(\zeta \otimes a) = \zeta^{\#i} \otimes \gamma_i(a)$$

if $i < p$. Now one checks easily that γ_p is well defined on $M \tilde{\otimes} A$. Since $1/2 \in R$ we see that $\gamma_p(\zeta \otimes a) = 0$ if $|\zeta \otimes a|$ is odd.

(2. 4) **Proposition.** *For modules of coefficients* M, N *in* Coef_R *and for an algebra* A *in* $\mathrm{Div\ alg}_R$ *there is a canonical isomorphism*

$$M \tilde{\otimes} (N \tilde{\otimes} A) \cong (M \tilde{\otimes} N) \tilde{\otimes} A$$

in $\mathrm{Div\ alg}_R$.

Proof. Both sides are well defined algebras in $\mathrm{Div\ alg}_R$, one checks that the product $\cup$ and the divided powers γ are the same on typical elements $\zeta \otimes \eta \otimes a$, $\zeta \in M$, $\eta \in N$, $a \in A$. //

Clearly we have the canonical isomorphism

(2. 5) $\underline{R} \tilde{\otimes} A = A.$

With respect to a monoid structure $\odot$ on coefficients M we now define the operation of M on A:

(2. 6) **Definition.** Let $(M, \odot)$ be a monoid in Coef_R and let A be a algebra in $\mathrm{Div\ alg}_R$. We say M *operates* on A if we have a morphism

$$M \tilde{\otimes} A \xrightarrow{\odot} A, \quad \zeta \otimes a \mapsto \zeta \odot a$$

in $\mathrm{Div\ alg}_R$ such that the diagrams

$$\begin{array}{ccc} M \tilde{\otimes} M \tilde{\otimes} A & \xrightarrow{1 \tilde{\otimes} \odot} & M \tilde{\otimes} A \\ \downarrow{\scriptstyle \odot \tilde{\otimes} 1} & & \downarrow{\scriptstyle \odot} \\ M \tilde{\otimes} A & \xrightarrow{\odot} & A \end{array}$$

$$\begin{array}{ccc} \underline{R} \tilde{\otimes} A & \xrightarrow{i \otimes 1} & M \tilde{\otimes} A \\ & \searrow{\scriptstyle =} & \downarrow{\scriptstyle \odot} \\ & & A \end{array}$$

commute.

(2. 7) **Proposition.** *With the identification in (1. 8) an* M-*algebra* A *with homotopy coefficients* M *in the sense of V (2. 9) is exactly given by an operation* $\odot$ *of* $(M, \odot)$ *on* A.

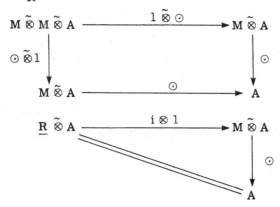

The category Div alg_M with homotopy coefficients $(M, \odot)$ is thus just the category of M-equivariant morphisms in Div alg_R. We have the forgetful functor

$$\text{Div alg}_M \to \text{Div alg}_R .$$

With respect to this functor we obtain:

(2. 8) Definition. Let A be an algebra in Div alg_R. We call an M-algebra A_M the M-extension of A if we have a morphism

$$i : A \to A_M \quad \text{in Div alg}_R$$

with the following universal property: For any M-algebra B and any $\phi : A \to B$ in Div alg_R there exists a unique M-equivariant ϕ_M so that

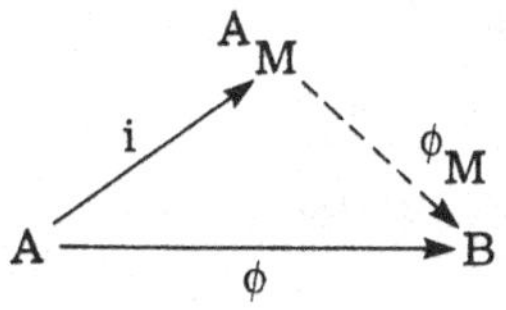

commutes in Div alg_R.

(2. 9) Theorem. For a monoid $(M, \odot)$ in Coef_R and for an algebra A in Div alg_R there exists the M-extension $A \to A_M$.

Proof. We define A_M as an object in Div alg_R by

$$A_M = M \tilde{\otimes} A .$$

Now M operates on $M \tilde{\otimes} A$ by

$$\odot = \odot \otimes 1_A : M \tilde{\otimes} M \tilde{\otimes} A \to M \tilde{\otimes} A$$

It is easily seen that

$$i = i \tilde{\otimes} 1_A : A = \underline{R} \tilde{\otimes} A \to M \tilde{\otimes} A$$

has the universal property in (2. 8). //

(2.10) **Definition.** Let $X = \{X_n, n \geq 0\}$ be a graded set with $X_0 = \emptyset$ empty. We call $i : X \to A_M(X)$ the free M-algebra generated by X if i is a function of degree 0 with the property: Any function of degree 0, $\phi : X \to A$, into an object of Div alg_M has a unique extension $\phi : A_M(X) \to A$ in Div alg_M with $\hat{\phi} \circ i = \phi$.

It is easy to see that $A_M(X)$ is the M-extension of the free object $A_R(X)$ in Div alg_R. Therefore we have:

(2.11) **Corollary.** There exist free objects in Div alg_M.

We know that for the graded set X with $X_0 = \emptyset$ the cohomology algebra

$$(2.12) \quad H^*\Big(\underset{x \in X}{\times} \Omega_R^{|x|}, R\Big) = A_R(X)$$

is the free algebra generated by X in Div alg_R, see V (1.3) and the proof of V (3.7).

§ 3. Extension of Lie algebras by homotopy coefficients

In V § 0 we defined the category Lie_R of Lie algebras over R. We now show that modules of coefficients M in Coef_R operate on this category too; we proceed in a similar way as in § 2 for algebras. We obtain an alternative description of the category Lie_M of Lie algebras over M, see V (2.12). As a main result we prove that for each Lie algebra L in Lie_R there exists a unique M-extension of L which is an object in Lie_M. The M-extension of a free object in Lie_R is a free object in Lie_M. Thus free objects exist in Lie_M.

Let R be a subring of $\mathbb{Q}$ with $1/2, 1/3 \in R$.

(3.1) **Definition.** For a Lie algebra $L = (L_*, [\ ,\ ])$ in Lie_R and for a module of coefficients in Coef_R we define the twisted product $L \tilde{\otimes} M$, which is again a Lie algebra in Lie_R. We set, as an abelian group,

$$(L \tilde{\otimes} M)_n = \text{FAG}\Big(\bigcup_{m \geq 1} L_m \times M^{m,n}\Big)/\sim$$

where $\text{FAG}(X)$ denotes the free abelian group generated by the set X. The relations are for $x, y \in L$, $\zeta, \eta \in M$.

(i) $(X, \xi) + (x, \eta) \sim (x, \xi + \eta)$

(ii) $(x, \xi) + (y, \xi) \sim (x + y, \xi) + \sum_{n\geq 2} (c_n(x, y), \gamma_n(\xi))$

(iii) $(\lambda x, \xi) \sim (x, \lambda\xi)$ for $\lambda \in R$.

The Zassenhaus term $c_n(x, y)$ is defined as in V (2.12). We denote with $x \tilde{\otimes} \xi$ the equivalence class of (x, ξ) in $L \tilde{\otimes} M$. Clearly elements of the form $x \tilde{\otimes} \xi$ generate $L \tilde{\otimes} M$ as an abelian group. The R-module structure on $L \tilde{\otimes} M$ is given by

(iv) $\lambda(x \tilde{\otimes} \xi) = (\lambda x) \tilde{\otimes} \xi = x \tilde{\otimes} (\lambda\xi)$, $\lambda \in R$.

The Lie bracket on $L \tilde{\otimes} M$ is defined on generators by

(v) $[x \tilde{\otimes} \xi, y \tilde{\otimes} \eta] = \sum_{M, N\geq 1} [x^M, y^N] \tilde{\otimes} (\gamma_M\xi \# \gamma_N\eta)$

(3.2) Proposition. $L \tilde{\otimes} M$ <u>has the well defined structure of a Lie algebra over</u> R.

Proof. Since γ_n is a homomorphism of R-modules by (1.1) (iii) we see that $L \tilde{\otimes} M$ is a well defined R-module. We now check that the Lie bracket is well defined on $L \tilde{\otimes} M$. Let

(1) $E_n = \bigcup_{m\geq 1} L_m \times M^{m,n}$

be the set of generators of $(L \tilde{\otimes} M)_n$. We have the canonical projection

(2) $\pi : FAG(E_n \times E_k) \to (L \tilde{\otimes} M)_n \otimes_R (L \tilde{\otimes} M)_k$

with $\pi((x, \xi), (y, \eta)) = (x \tilde{\otimes} \xi) \otimes (y \tilde{\otimes} \eta)$. The relations for π are:

(3)
$A, B \in FAG(E_n)$, $C, D \in FAG(E_k)$

$A \sim B \Rightarrow (A, c) \sim (B, c)$ for all $c \in E_k$

$C \sim D \Rightarrow (a, C) \sim (a, D)$ for all $a \in E_n$

$((\lambda x, \xi), (y, \eta)) \sim ((x, \xi), (\lambda y, \eta))$ for $\lambda \in R$.

We have to prove that the homomorphism

(4) $\quad K : FAG(E_n \times E_k) \to (L \tilde{\otimes} M)_{n+k}$

defined on generators by

$$K((x, \zeta), (y, \eta)) = \sum_{M, N \geq 1} [x^M, y^N] \tilde{\otimes} (\gamma_M \zeta \# \gamma_N \eta)$$

factors over π. We do this for the relation $(A, c) \sim (B, c)$ where $A = (x, \zeta) + (x', \zeta) \sim B = \sum_{n \geq 1} (c_n(x, x'), \gamma_N(\zeta))$ and $c = (y, \eta) \in F_k$. We have

(5)
$$\begin{aligned} K(A, c) &= K(((x, \zeta), c) + ((x', \zeta), c)) \\ &= \sum_{M, N \geq 1} ([x^M, y^N] \tilde{\otimes} (\gamma_M \zeta \# \gamma_N \eta) + [x'^M, y^N] \tilde{\otimes} (\gamma_M \zeta \# \gamma_N \eta)) \\ &= \sum_{M, N \geq 1} ([x^M, y^N] + [x'^M, y^N]) \tilde{\otimes} (\gamma_M \zeta \# \gamma_N \eta) \end{aligned}$$

The last equation follows from (ii) (3.1) since γ_j vanishes on products $\gamma_M \zeta \# \gamma_N \eta$ for $j \geq 2$, see (1.1) (e).

On the other hand we have

(6)
$$K(B, c) = (\sum_{n \geq 1} \sum_{m, N \geq 1} [c_n(x, x')^m, y^N]) \tilde{\otimes} (\gamma_m \gamma_n(\zeta)) \# \gamma_N \eta$$

We now fix a prime $p \geq 3$. Then $K(A, c) = K(B, c)$ follows from the equation mod p

(7)
$$[x^M, y^N] + [x'^M, y^N] \equiv \sum_{\substack{m \cdot n = M \\ m, n \geq 1}} \varepsilon [c_n(x, x')^m, y^N]$$

where $\varepsilon = (m \cdot n)! / (m!)^n (n!)$. In (7) only indices M, N, m, n appear which are powers of the prime p. In this case we know for the universal enveloping of L that (mod p)

(8)
$$[x^M, y^N] \equiv [x^{\otimes M}, y^{\otimes N}], \quad \text{see I § 4,}$$

where $x^{\otimes M} = x \cdot \ldots \cdot x$ is the M-fold product in UL. Therefore (7) follows from the equation

(9)
$$x^{\otimes M} + x'^{\otimes M} = \sum_{m \cdot n = M} \varepsilon\, c_n(x, x')^{\otimes m}$$

which we already described in I § 4.

For relations $A \sim B$ of type (i) or (iii) it is easy to check that $K(A, c) = K(B, c)$. Similarly we treat the relations $C \sim D$ in (3). This proves that the bracket is well defined on $L \tilde{\otimes} M$. We still have to check that it is a Lie-bracket, see V (0.4). It is enough to consider anti symmetry and the Jacobi identity on generators: These equations follow easily from (8) and (V) in (3.1). //

(3.3) **Proposition.** For modules of coefficients M, N in $\text{Coef}_{\underline{R}}$ and for a Lie algebra L in $\text{Lie}_{\underline{R}}$ there is a canonical isomorphism

$$(L \tilde{\otimes} N) \tilde{\otimes} M \cong L \tilde{\otimes} (N \tilde{\otimes} M)$$

in $\text{Lie}_{\underline{R}}$.

Proof. The isomorphism maps a generator $(x \tilde{\otimes} \zeta) \tilde{\otimes} \zeta'$ to the generator $x \tilde{\otimes} (\zeta \tilde{\otimes} \zeta')$. For the proof that this map is well defined we consider the R-module

$$\mathcal{L} = (L \tilde{\otimes} N) \tilde{\otimes}_N (N \tilde{\otimes} M)$$

with

$$\mathcal{L}_n = \text{FAG}\Big(\sum_{m \geq 1} (L \tilde{\otimes} N)_m \times (N \tilde{\otimes} M)^{m, n} \Big) / \sim$$

The relations are $(x, y \in L \tilde{\otimes} N,\ \zeta, \eta \in N \tilde{\otimes} M)$

(i) $(x, \zeta) + (x, \eta) \sim (x, \zeta + \eta)$

(ii) $(x, \zeta) + (y, \zeta) \sim (x + y, \zeta) + \sum_{n \geq 2} (c_n(x, y), \gamma_n(\zeta))$

(iii) $(x \tilde{\otimes} \lambda, \zeta) \sim (x, \lambda \tilde{\otimes} \zeta)$ for $\lambda \in N$.

From the retraction

$$\underline{R} \overset{i}{\hookrightarrow} N \overset{\varepsilon}{\to} \underline{R}$$

we obtain well defined maps

$$(L \tilde{\otimes} N) \tilde{\otimes}_{\underline{R}} (\underline{R} \tilde{\otimes} M) \underset{\varepsilon}{\overset{i}{\rightleftarrows}} \mathcal{L}$$

$$(L \tilde{\otimes} \underline{R}) \otimes_{\underline{R}} (N \tilde{\otimes} M) \underset{\varepsilon'}{\overset{i'}{\rightleftarrows}} \mathcal{L}$$

with $\varepsilon i = 1$, $\varepsilon' i' = 1$. This proves that i and i' are injective. On the other hand it is easy to see that i and i' are surjective. Therefore $i'^{-1} i$ is a well defined isomorphism of R-modules. Since by (V) (3.1)

$$[x \tilde{\otimes} \zeta^M, y \tilde{\otimes} \eta^N] = \sum_{\substack{n \geq N \\ m \geq M}} [x^n, y^m] \tilde{\otimes} \Gamma_n^N(\zeta) \# \Gamma_m^M(\eta).$$

Compare V (2.3) (i), we find that $i'^{-1} i$ is in fact an isomorphism of Lie algebras. //

(3.4) **Remark.** If the invariants γ_r for $r \geq 2$ are trivial in M the relations (3.1) show that we have a canonical isomorphism

$$L \tilde{\otimes} M \cong L \otimes M$$

of R-modules, see V (3.12). We therefore consider γ as the twisting of $L \tilde{\otimes} M$.

Clearly we have the canonical isomorphism

(3.5) $\quad L \tilde{\otimes} \underline{R} \cong L$

which we already used in the proof of (3.3). Similarly as in (2.6) we now define with respect to a monoid-structure on coefficients M an operation of M on a Lie algebra L.

(3.6) **Definition.** Let $(M, \odot)$ be a monoid in Coef_R and let L be a Lie algebra in Lie_R. We say M operates on L if we have a morphism

$$L \tilde{\otimes} M \xrightarrow{\odot} L, \quad x \tilde{\otimes} a \mapsto x \odot a$$

in Lie_R such that the diagrams

$$\begin{array}{ccc} L \tilde{\otimes} M \tilde{\otimes} M & \xrightarrow{1 \tilde{\otimes} \odot} & L \tilde{\otimes} M \\ {\scriptstyle \odot \tilde{\otimes} 1} \downarrow & & \downarrow {\scriptstyle \odot} \\ L \tilde{\otimes} M & \xrightarrow{\odot} & L \end{array}$$

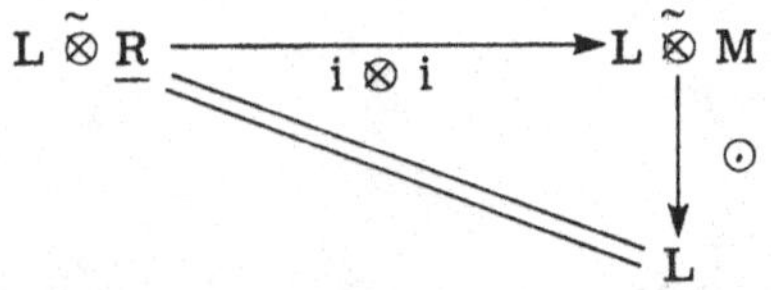

commute.

Now we have:

(3.7) **Proposition.** With the identification in (1.8) a Lie algebra L over M in Lie_M defined in V (2.12) is exactly given by an operation ⊙ of (M, ⊙) on L.

Therefore the category Lie_M with homotopy coefficients (M, ⊙) is thus just the category of M-equivariant morphisms in Lie_R. We consider the forgetful functor

$$Lie_M \to Lie_R$$

which leads to the following definition (dual to (2.8)):

(3.8) Definition. Let L be a Lie algebra in Lie_R. We call a Lie algebra L_M in Lie_M the M-extension of L if we have a morphism

$$i : L \to L_M \quad \text{in } Lie_R$$

with the universal property: For any Lie algebra K in Lie_M and any $\phi : L \to K$ in Lie_R there exists a unique M-equivariant ϕ_M so that

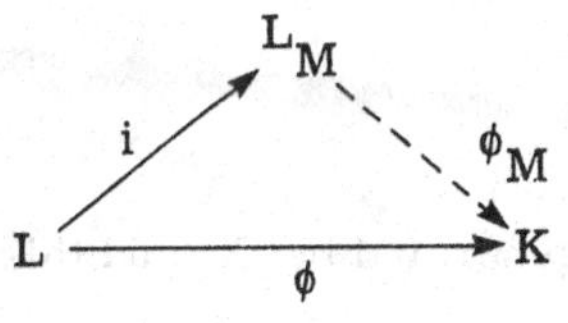

commutes in Lie_R.

(3.9) Theorem. For a monoid (M, ⊙) in $Coef_R$ and for a Lie algebra L in Lie_R there exists the M-extension $L \to L_M$.

Proof. We define L_M as an object in Lie_R by

$$L_M = L \tilde{\otimes} M$$

Now M operates on $L \tilde{\otimes} M$ by

$$\odot = 1_L \tilde{\otimes} \odot : L \tilde{\otimes} M \tilde{\otimes} M \to L \tilde{\otimes} M .$$

It is easily checked that

$$i = 1_L \tilde{\otimes} i : L = L \tilde{\otimes} \underline{R} \to L \tilde{\otimes} M$$

has the universal property in (3.8). //

(3.10) **Definition.** Let $X = \{X_n, n \geq 0\}$ be a graded set with $X_0 = \emptyset$. We call $i : X \to L_M(X)$ the <u>free</u> Lie algebra in Lie_M generated by X if i is a function of degree 0 with the property: Any function $\phi : X \to K$ into an object of Lie_M of degree 0 has a unique extension $\hat{\phi} : L_M(X) \to K$ in Lie_M with $\hat{\phi} \circ i = \phi$.

Clearly $L_M(X)$ is the M-extension of the free object $L_R(X)$ in Lie_R. Therefore we have

(3.11) **Corollary.** <u>There exist free objects in</u> Lie_M.

We know that for the graded set X with $X_0 = \emptyset$ the homology Lie algebra

$$(3.12) \quad PH_*(\Omega \bigvee_{x \in X} \Sigma S_R^{|x|}, R) = L_R(X)$$

is the free Lie algebra in Lie_R generated by X. In fact

$$H_*(\Omega \bigvee_{x \in X} \Sigma S_R^{|x|}, R)$$

is the primitively generated tensor algebra generated by X and this algebra is free as an R-module.

§ 4. The Hilton Milnor theorem and its dual

Let R be a subring of $\mathbb{Q}$ with $1/2, 1/3 \in R$ and let $M = M_R^{**}$ be the module of spherical homotopy coefficients as defined in V § 2. In chapter V and chapter VI we studied suspensions ΣX and loop spaces

ΩY which are decomposable and even splittable, see VI (3. 3).

In the following let X and ΩY be connected and of finite type over R.

(4. 1) Theorem. If ΣX is splittable a right inverse $\sigma : H^*(X, R) \to M^*(X, R)$ of the degree map deg is an M-extension of the algebra $H^*(X, R)$. That is, σ yields an isomorphism in Div alg_M

$$M^*(X, R) \cong M \tilde{\otimes} H^*(X, R)$$

The dual of this theorem is:

(4. 2) Theorem. If ΩY is splittable a right inverse $\tau : PH_*(\Omega Y, R) \to \pi_*(\Omega Y) \otimes R$ of the Hurewicz map ϕ is an M-extension, that is, τ induces an isomorphism in Lie_M

$$\pi_*(\Omega Y) \otimes R \cong PH_*(\Omega Y, R) \tilde{\otimes} M .$$

Proof. (4. 1) and (4. 2) are in fact special cases of VI (3. 4). By VI (3. 4) we know (n even)

$$\begin{aligned} M^n(X, R) = [\Sigma X_R, \Sigma S^n_R] &= \exp_{\underline{R}}(H^*(X, R), \pi_*\Omega^n_R) \\ &= (M \tilde{\otimes} H^*(X, R))_n . \end{aligned}$$

The last equation is true by VI (2. 1) since the Lie bracket in $\pi_*\Omega^n_R$ is trivial.

Moreover we know from VI (3. 4)

$$\begin{aligned} \pi_n(\Omega Y) \otimes R = [\Sigma S^n_R, Y_R] &= \exp_{\underline{R}}(M^*(S^n_R, R), PH_*(\Omega Y, R)) \\ &= (PH_*(\Omega Y, R) \tilde{\otimes} M)_n \end{aligned}$$

For the last equation compare the relations in VI (2. 1) and in (3. 1). //

In V (1. 3) we defined the universal objects Ω^n_R for the spherical cohomotopy algebra $M^*(\cdot , R)$. For these universal objects we have by (2. 12):

(4.3) Corollary. *Let* X *be a graded set with* X_n *finite for* $n \geq 0$ *and* X_0 *empty. Then*

$$M^*\big(\underset{x\in X}{\times}\ \Omega_R^{|x|},\ R\big) \cong M \tilde{\otimes} A_R(X)$$

is isomorphic in Div alg_M *to the free* M-*algebra generated by* X.

This corollary is dual to the following Hilton-Milnor theorem [23]. We point out that we need no basic commutators to formulate it.

The spheres ΣS_R^n are the universal objects for the homotopy Lie algebras $\pi_*(\Omega\ .\ ) \otimes R$. We have by (3.12):

(4.4) Corollary. *Let* X *be a graded set with* X_n *finite for* $n \geq 0$ *and* X_0 *empty. Then*

$$\pi_*\big(\Omega \underset{x\in X}{\vee}\ \Sigma S^{|x|}\big) \otimes R \cong L_R(X) \tilde{\otimes} M$$

is isomorphic in Lie_M *to the free Lie algebra over* M *generated by* X.

(4.3) and (4.4) are consequences of (4.1) and (4.2) and of the following result:

(4.5) Theorem. (A) *If* ΣX *and* ΣY *are splittable then so is* $\Sigma(X \times Y)$.

(B) *If* ΩX *and* ΩY *are splittable then so is* $\Omega(X \vee Y)$.

Proof. The proof is purely formal and relies on the existence of free products $\coprod$ in Div alg_R and Lie_R. Let σ_X and σ_Y be splittings for ΣX and ΣY respectively. Then

$$\begin{array}{ccc}
H^*(X\times Y,\ R) \cong H^*(X) \coprod H^*(Y) & & \\
 & \searrow & \downarrow \sigma_X \coprod \sigma_Y \\
\sigma_{X\times Y} & & M^*(X,\ R) \coprod M^*(Y,\ R) \\
 & & \downarrow (p_1^*,\ p_2^*) \\
 & & M^*(X \times Y,\ R)
\end{array}$$

is a splitting for $\Sigma(X \times Y)$, p_1 and p_2 are the projections of $X \times Y$ onto X and Y respectively.

Similarly we obtain the result for $\Omega(X \vee Y)$. Let τ_X and τ_Y be splittings for ΩX and ΩY respectively. Then

$$\begin{array}{ccc}
PH_*(\Omega(X \vee Y)) \cong & PH_*(\Omega X) \coprod PH_*(\Omega Y) \\
 & \downarrow \tau_X \coprod \tau_Y \\
\tau_{X \vee Y} \searrow & \pi_*(\Omega X_R) \coprod \pi_*(\Omega Y_R) \\
 & \downarrow (i_{1*}, i_{2*}) \\
 & \pi_*(\Omega(X_R \vee Y_R))
\end{array}$$

is a splitting for $\Omega(X \vee Y)$. i_1 and i_2 are the inclusions of X and Y into $X \vee Y$ respectively. //

As a special case of VI (3.4) we now have as a consequence of (4.1) and (4.2):

(4.6) **Corollary.** *Let* X *and* ΩY *be connected and of finite type over* R. *If* ΣX *and* ΩY *are splittable the group* $[\Sigma X, Y]_R$ *depends only on the cohomology algebra* $H^*(X, R)$ *and the homology Lie algebra* $PH_*(\Omega Y, R)$. *In fact we have the isomorphism of R-local groups*

$$[\Sigma X, Y]_R \cong \exp_M(M \tilde{\otimes} H^*(X, R), PH_*(\Omega Y, R) \tilde{\otimes} M)$$

where M *is the module of spherical homotopy coefficients.*

In particular for graded sets X, Y with X_n, Y_n finite and $X_0 = Y_0 = \emptyset$ we have the isomorphism of R-local groups

$$[\Sigma \times_{x \in X} \Omega_R^{|x|}, \bigvee_{y \in Y} \Sigma S_R^{|y|}] \cong \exp_M(M \tilde{\otimes} A_R(X), L_R(Y) \tilde{\otimes} M)$$

A further application is the algebraic characterization of the most simple homotopy categories:

(4.7) **Definition.** We call X a Σ_1-space and Y an Ω_1-space if X and Y are homotopy equivalent to

$$\bigvee_{a \in A} \Sigma S_R^{|a|} \quad \text{and} \quad \times_{a \in A} \Omega_R^{|a|}$$

respectively, where A is a graded set with A_n finite and A_0 empty. Let $\mathrm{Top}\,\Sigma_1$ respectively $\mathrm{Top}\,\Omega_1$ be the homotopy categories of Σ_1- and Ω_1-spaces.

(4.8) **Theorem.** <u>We have equivalences of categories</u>

$$\pi_*(\Omega\,.)\otimes R : \mathrm{Top}\,\Sigma_1 \to F\,\mathrm{Lie}_M$$

$$M^*(\,.\,,R) \quad : \mathrm{Top}\,\Omega_1 \to F\,\mathrm{Div\,alg}_M$$

<u>where</u> F <u>denotes the sub-category of free objects of finite type over</u> R.

This result allows us to say that the structure of spherical homotopy coefficients M as a monoid in Coef_R is the <u>primary structure</u> of homotopy groups of spheres. This primary structure determines the homotopy categories $\mathrm{Top}\,\Sigma_1$ and $\mathrm{Top}\,\Omega_1$.

In fact let Coef_R^1 be the category of monoids in Coef_R. Then each object M in Coef_R determines the categories

$$\Sigma M = F\,\mathrm{Lie}_M$$

$$\Omega M = F\,\mathrm{Div\,alg}_M$$

in such a way that for the spherical homotopy coefficients M we have equivalences of categories

$$\Sigma M \sim \mathrm{Top}\,\Sigma_1$$

$$\Omega M \sim \mathrm{Top}\,\Omega_1\ .$$

This leads to the following type of problems:

(4.9) **Definition.** A Σ_2-space is the mapping cone of a map in $\mathrm{Top}\,\Sigma_1$ and a Ω_2-space is the fibre of a map in $\mathrm{Top}\,\Omega_1$. Let $\mathrm{Top}\,\Sigma_2$, respectively $\mathrm{Top}\,\Omega_2$, be the homotopy category of these spaces.

(4.10) **Problem.** Introduce <u>secondary structure</u> on an object in Coef_R^1 such that the spherical homotopy coefficients have this structure. Let Coef_R^2 be the category of objects in Coef_R^1 with this additional structure.

Introduce the structure in such a way that each object $\hat{M}$ in in Coef_R^2 determines purely algebraically categories $\Sigma\hat{M}$ and $\Omega\hat{M}$ so that for the spherical $\hat{M}$ there are equivalences of categories

$$\Sigma\hat{M} \sim \mathrm{Top}\,\Sigma_2, \quad \Omega\hat{M} \sim \mathrm{Top}\,\Omega_2 .$$

LITFRATURF

1. D. W. Anderson. The e-invariant and the Hopf invariant. Top. 9 (1970), 49-54.

2. M. Arkowitz and C. R. Curjel. Groups of homotopy classes. Lect. Notes in Math. 4 (1964), Springer Verlag, Berlin, Heidelberg, New York.

3. M. Arkowitz and C. R. Curjel. On the number of multiplications of an H-space. Top. 2 (1963), 205-8.

4. M. Arkowitz and C. R. Curjel. On maps of H-spaces. Top. 6 (1967), 137-48.

5. W. Barcus and M. G. Barratt. On the homotopy classification of the extensions of a fixed map. Trans. Amer. Math. Soc. 88 (1958), 57-74.

6. M. G. Barratt. Higher Hopf invariants (mimeographed notes) University of Chicago (Summer 1957).

7. H. J. Baues. Relationen für primäre Homotopieoperationen und eine verallgemeinerte FHP Sequenz. Ann. Scient. de l'Fcole Normale Superieure fasc. 4, 8 (1975), 509-33.

8. H. J. Baues. Hopf invariants for reduced products of spheres. Proceed. AMS 59, 1 (1976), 169-74.

9. H. J. Baues. Rationale Homotopietypen, manuscr. math. 20, (1977), 119-31.

10. H. J. Baues and J. M. Lemaire. Minimal models in homotopy theory. Math. Ann. 225 (1977), 219-42.

11. H. J. Baues. Obstruction theory. Lect. Notes in Math. 628 (1977). Springer Verlag, Berlin, Heidelberg, New York.

12. G. Baumslag. Lecture Notes on Nilpotent groups. AMS Regional Conference Series No. 2 (1971).

13. J. M. Boardman and B. Steer. On Hopf invariants. Comment. Math. Helv. 42 (1967), 180-221.

14. B. Cenkl and R. Porter. Malcev's completion of a group and differential forms. (Preprint) Northeastern Boston (1979).

15. F. R. Cohen and J. C. Moore and J. A. Neisendorfer. Torsion in homotopy groups. Ann. Math. 109 (1979), 121-68.

16. G. F. Cooke and L. Smith. Mop p decompositions of Co-H-spaces and applications. Math. Z. 157 (1977), 155-77.

17. Y. Felix. Classification homotopique des espaces rationnels a cohomologie donée. Thése 1979, Louvain la Neuve.

18. T. Ganea. Co-groups and suspensions. Inventiones math. 9 (1970), 185-97.

19. B. Gray. On the sphere of origin of infinite families in the homotopy groups of spheres. Top. 8 (1969), 219-32.

20. A. Haefliger. Rational homotopy of the space of sections of a nilpotent bundle. Preprint.

21. S. Halperin and J. Stasheff. Obstructions to homotopy equivalences. Adv. math. (1979).

22. S. Halperin and C. Watkiss. Preprint.

23. P. J. Hilton. On the homotopy groups of the union of spheres. J. London Math. Soc. 30 (1955), 154-72.

24. P. J. Hilton. Note on quasi Lie rings. Fund. Math. 43 (2) (1957), 230-7.

25. P. J. Hilton and C. Mislin and J. Roitberg. Localization of nilpotent groups and spaces. North Holland Math. Studies 15, North Holland Company, Amsterdam (1975).

26. N. Jacobson. Lie algebras. Interscience tracts in Pure and Appl. Math. vol. 10 (1962), New York.

27. I. M. James. Reduced product spaces. Ann. Math. 62 (1955), 170-97.

28. L. L. Larmore and F. Thomas. Mappings into loop spaces and central group extensions. Math. S. 128 (1972), 277-96.

29. M. Lazard. Sur les groupes nilpotents et les anneaux de Lie. Ann. Sci. Fcole Norm. Sup. (3) 71 (1954), 101-90.

30. W. Magnus, A. Karrass and D. Solitar. Combinatorial group theory. Pure and Appl. Math. vol 13, Interscience, New York (1966).

31. A. I. Malcev. On a class of homogeneous spaces. Izv. Akad. Nauk. SSSR Ser. Mat. 13 (1949), 9-32; Fnglish Translation, Amer. Math. Soc. Transl. Series 1, 39 (1962).

32. F. Y. Miller. De Rham Cohomology with arbitrary coefficients. Top. 17 (1978), 193-203.

33. J. W. Milnor and J. C. Moore. On the structure of Hopf algebras. Ann. of Math. 81 (1965), 211-264.

34. D. Quillen. Rational homotopy theory. Ann. of Math. 90 (1966), 205-95.

35. H. Scheerer. Gruppen von Abbildungen in Produkten von Filenberg MacLane Räumen. Math. Ann. 210 (1974), 281-94.

36. J. P. Serre. Homologie singulaire des espaces fibrés. Ann. Math. 54 (1951), 425-505.

37. F. F. A. da Silveira. Homotopie rationelle d'espaces fibrés Thèse No. 1918 Genève (1979).

38. H. Spanier. Algebraic topology. MacGraw Hill (1966).

39. B. Steer. Generalized Whitehead products. Quart. J. Math. Oxford (2), 14 (1963), 29-40.

40. D. Sullivan. Infinitesimal computations in topology. Publ. de I. H. E. S. 47 (1978), 269-331.

41. F. Thomas. The generalized Pontrjagin cohomology operations and rings with divided powers. Memoirs AMS 27, (1957).

42. M. Vigué. Quelques problèmes d'homotopie rationelle. Thèse (1978), Lille.

43. G. W. Whitehead. On mappings into group like spaces. Commentarii Math. Helv. 21 (1954), 320-8.

44. H. Zassenhaus. Über Lie'sche Ringe mit Primzahlcharakteristik. Abhandlungen Math. Sem. Hansische Uni. 13 (1940), 1-100.

INDEX

admissible ordering 16
algebra (graded commutative) 102
augmentation 114

Baker-Campbell-Hausdorff formula 15
Barcus-Barratt formula 51
basic commutators 46
base-point 36

category of connected Lie algebras 102
  algebras with divided powers 103
  M-algebras 112
  M-Lie algebras 112
  coefficients 135
classifying space 37
coalgebra 27, 123
codimension 39
coformal type 133
cohomotopy algebra 104
co-H-space, Co-H-map 36, 43, 131
commutator 17, 40
cross product 114
cup product 41

decomposition 63
decomposable 117
degree map 113, 115
desuspension problem 133
dimension 39
divided powers 103

e-invariant 93
exponential function 15
  commutator 22
  group 27, 124
exterior cup product 40

formal type 133
free Lie algebra 15, 17
  non-associative algebraic object 17
  group 17
  monoid 30
  abelian group 30
  M-algebra 144
  M-Lie algebra 150

graded 14

Hilton-Milnor theorem 46
Hilton-Hopf invariant 46
homotopy Lie algebra 103, 104
homotopy coefficients
Hopf algebra 114
H-space 37
H-maps 131
Hurewicz map 113, 114

integral Lie element 15
inverse limit 105, 123
iterated brackets 39
iterated Whitehead products 38, 39

Jacobi identity 38
James-Hopf invariant 43, 91
James filtration 133, 134
J-homomorphism 94

left distributivity law 44
length of a product 39
lexicographical ordering 17
Lie algebra 102, 122
Lie bracket 14, 15, 122
local group 101
loop space 36

Malcev completion 27
M-algebra 111
M-extension of algebras 143
M-extension of Lie algebras 149
Milnor-Moore theorem 115
module of homotopy coefficients 109
  spherical homotopy coefficients 110
monoid in the category of coefficients 137
monotone 16

nilpotent 26, 101, 122

partition 23
primary structure 154

rational group 26, 101
rationalization 26
reduced product 42

Samelson product 38
secondary structure 154
space 36
spherical 68
splittable 127
suspension 36

tensor algebra 14
tensor product of coefficients 136
twisted product 139, 144
twisting 148

uniquely divisible 26
universal enveloping algebra 114

weight 15
Whitehead product 37
Witt-Hall identity 18, 19

Zassenhaus formula 15, 29
Zassenhaus term (general) 31